Cliffs Quick Review

P9-CMQ-392

Statistics

by
David H. Voelker, M.A. and Peter Z Orton, Ed.M.

Series Editor
Jerry Bobrow, Ph.D.

Cliffs Notes
INCORPORATED
LINCOLN, NEBRASKA 68501

FIRST EDITION

ISBN 0-8220-5349-7

CONTENTS

CONTENTS

CONTENTS

CONTENTS

Students often find statistics confusing and intimidating. It doesn't have to be. In this small volume, we have tried to make this important reasoning process clear and concise. The text can serve as an introduction to the field or as a helpful supplement and review. We hope that it serves you well.

The authors

What exactly is **statistics?**

Types of Statistics

For many people, statistics means numbers—numerical facts, figures, or information. Reports of industry production, baseball batting averages, government deficits, and so forth, are often called statistics. To be precise, these numbers are **descriptive statistics** because they are numerical data that describe phenomena. Descriptive statistics are as simple as the number of children in each family along a city block or as complex as the annual report released from the U.S. treasury department.

In these pages, we are concerned with two ways of representing descriptive statistics: numerical and pictorial.

Numerical statistics. Numerical statistics are numbers, but clearly, some numbers are more meaningful than others. For example, if you are offered a purchase price of $1 for an automobile on the condition that you also buy a second automobile, the price of the second automobile will be a major consideration (its price could be $1,000,000 or only $1,000), and thus, the average—or **mean**—of the two prices will be the important statistic. The different meaningful numerical statistics are discussed in Numerical Measures (page 27).

Pictorial statistics. Taking numerical data and presenting it in pictures or graphs is what is known as pictorial statistics. Showing data in the form of a graphic can make complex and confusing information appear more simple and straightforward. The more commonly used graphics in statistics are discussed in Graphic Displays (page 9).

Method of Statistical Inference

Statistics is also a method, a way of working with numbers to answer puzzling questions about both human and non-human phenomena. Questions answerable by using the "method" of statistics are many and varied: Which of several techniques is best for teaching reading to third-graders? Will a new medicine be more effective than the old one? Can you expect it to rain tomorrow? What's the probable outcome of the next presidential election? Which assembly-line process produces fewer faulty carburetors? How can a polling organization make an accurate prediction of a national election by questioning only a few thousand voters? And so on.

For our purposes, statistics is both a collection of numbers and/or pictures *and* a process: the art and science of making accurate guesses about outcomes involving numbers.

So, fundamentally, the goals of statistics are

1. to describe sets of numbers, and
2. to make accurate inferences about groups based upon incomplete information

Steps in the Process

Making accurate guesses requires groundwork. The statistician must:

1. gather **data** (numerical information); then,
2. organize it (sometimes in a pictorial); then,
3. analyze it (using tests of significance, and so forth)

in order to make an educated, better-than-chance hunch. In this book, we shall show you how to do precisely these procedures and finally how to use your analysis to draw an **inference**—an educated statistical guess—to solve a particular problem. While these steps may appear simple and indeed some of them are, sound statistical method requires that you perform them in certain prescribed ways. Those ways are the heart and soul of statistics.

Making Predictions

For example, suppose that you decide to sell commemorative t-shirts at your town's centennial picnic. You know that you can make a tidy profit, but only if you can sell most of your supply of shirts because your supplier won't buy any of them back. How many shirts can you reasonably plan on selling?

Your first question, of course, is: How many people will be attending the picnic? Suppose you know that 100,000 tickets have been sold to the event. How many t-shirts should you purchase from your supplier to sell on the day of the event? 10,000? 50,000? 100,000? 150,000? How many of each size—small, medium, large, extra large? And the important question: How many must you sell in order to make some profit for your time and effort and *not* be left with an inventory of thousands of unsold t-shirts?

Ideally you need to have an accurate idea—*before* you buy your inventory of t-shirts—just how many ticket holders will want to purchase centennial t-shirts and which sizes they will want. But, obviously, you have neither the time nor the resources to ask each of those 100,000 people if they plan to purchase a commemorative t-shirt. If, however, you could locate a small number of those ticket holders—for example, 100—and get an accurate count of how many of those 100 would purchase a t-shirt, you'd have a better idea of how many of the 100,000 ticket holders would be willing to buy one.

That is, of course, if the 100 ticket holders that you asked (called the **sample**) are not too different in their intentions to purchase a t-shirt from the total 100,000 ticket holders (called the **population**). If the sample is indeed representative (typical) of the population, you could expect about the same percentage of t-shirt sales (and sizes) for the population as for the sample, all things being equal. So if 50 of your sample of 100 people say they can't wait to plunk down $10 for a centennial t-shirt, it would be reasonable to expect that you would sell about 50,000 t-shirts to your population of 100,000. (At just $1 profit per t-shirt, that's $50,000!)

But before you start shopping to buy a yacht with your profits, remember that this prediction of total t-shirt sales relies heavily upon the sample being representative (similar to the population), which may not necessarily be the case with your sample. You may have inadvertently selected a sample that has more expendable income, or a greater proportion of souvenir t-shirt enthusiasts, or who knows what else. Are you reasonably certain that the intentions of the sample of 100 ticket holders reflect the intentions of the 100,000 ticket holders? If not, you may quite possibly be stuck with tens of thousands of centennial t-shirts and no profit to splurge on a yacht.

You can see why choosing a random sample is a critical part of the process of statistics, and why sampling is discussed in this book (page 55).

Comparing Results

Making predictions is only one use of statistics. Let's suppose you have recently developed a new headache/pain remedy that you call Ache-Away. Should you produce Ache-Away in quantity and make it available to the public? That would depend, among other concerns, upon whether Ache-Away is more effective than the old remedy. How can you determine that?

One way might be to administer both remedies to two separate groups of people, collect data on the results, and then statistically analyze that data to determine if Ache-Away is more effective than the old remedy. And what if the results of this test showed Ache-Away to be more effective? How certain can you be that *this* particular test administration is indicative of *all* tests of these two remedies? Perhaps the group taking the old remedy (the control group) and the group taking Ache-Away (the treatment group) were so dissimilar that the results were due *not* to the pain remedies but to the differences between the groups.

It's possible that the results of this test are far off the results that you would get if you tried the test several more times. You certainly do not want to foist a questionable drug upon an unsuspecting public based upon untypical test results. How certain can you be that you can put your faith in the results of your tests? You can see that the problems of comparing headache remedy results can produce headaches of their own.

This type of statistical procedure is discussed in Principles of Testing and Univariate Inferential Tests (pages 69 and 85), respectively).

Probability

One of the most familiar uses of statistics is to determine the chance of some occurrence. For instance, what are the chances that it will rain tomorrow or that the Boston Red Sox will win a World Series? These kinds of probabilities, although interesting, are not the variety under discussion here. Rather, we are examining the probability in statistics that deals with classic theory and frequency theory—events that can be repeated over and over again, independently, and under the same conditions.

Coin tossing and card drawing are two such examples. A *fair* coin (one that is not weighted or *fixed*) has an equal chance of landing heads as landing tails. A typical deck of cards has 52 different cards—13 of each suit (hearts, clubs, diamonds, and spades)—and each card or suit has an equal chance of being drawn. This kind of event forms the basis of your understanding of probability and enables you to find solutions to everyday problems that seem far removed from coin tossing or card drawing.

Probability is discussed beginning on page 41.

Common Statistics Mistakes

Whether you are new to the study of statistics or have been using statistical analysis for years, there is a likelihood that you will make (or already have made) errors in your application of principles explained in this book. In fact, some particular errors are committed so frequently in statistical analysis that we feel obliged to call them to your attention in the hope that by being prepared for them, you will avoid their pitfalls.

Pie charts, bar charts, and histograms are graphic displays of numerical data—simple collections of numbers made more understandable by use of visuals. This section discusses the more common graphic displays used in statistical analysis.

As an example, consider the yearly expenditures of a college undergraduate. After collecting her *data* (expense records) for the past year, she finds the following expenditures:

Tuition fees	$5,000
Room and board	9,000
Books and lab	2,000
Clothes/cleaning	1,000
Transportation	2,000
Insurance and miscellaneous	1,000

These figures, although presented in categories, do not allow for easy analysis. The reader must expend extra effort in order to compare amounts spent or relate individual proportions to the total. For ease of analysis, these data can be presented pictorially.

Bar Chart

One way to pictorially display these numbers is with a **bar chart:**

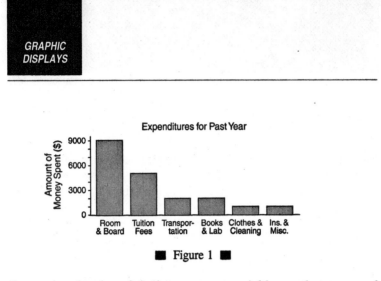

■ Figure 1 ■

Comparing the size of the bars, you can quickly see that room and board expenses are nearly double tuition fees, and tuition fees are more than double books and lab or transportation expenses.

A bar chart may also be placed on its side with the bars going horizontally:

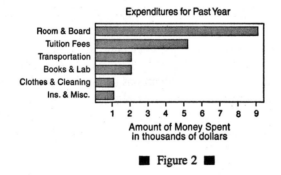

■ Figure 2 ■

In each bar chart, vertical or horizontal, the amounts are ordered from highest to lowest or lowest to highest, making the chart clearer and more easily understood. Space is left between each of the bars in order to define the categories as being different.

The bottom line in Figure 1 and the left-hand side in Figure 2 indicate 0. Although typical, this presentation need not always be used. Finally, while the lengths of the bars may be different, their *thicknesses* are the same.

Dot Plot

Dot plots are similar to bar graphs. Typically used for a small set of values, a dot plot uses a *dot* for each unit of measurement; thus, the preceding undergraduate expense data represented in a dot plot could look like this:

■ Figure 3 ■

Pie Chart

The previous charts have a limitation: it's difficult to see what portion of the total each item comprises. If knowing about a "part of the whole" is an important consideration, then a **pie chart** is a better choice for showing the same data. A pie chart may also display each category's percentage of the total. Using the same data from the undergraduate expenditures above:

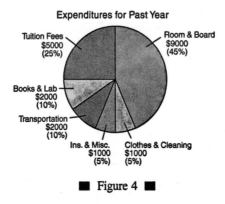

Expenditures for Past Year

■ Figure 4 ■

The parts of the circle (or *pie*) match, in size, each category's percentage of the total. The parts of the pie chart are ordered from highest to lowest for easier interpretation of the data. Pie charts work best with only a few categories; too many categories will make a pie chart confusing.

Ogive

Data may be expressed using a single line. An **ogive** (a cumulative line graph) is best used when you wish to display the total at any given time. The relative slopes from point to point will indicate greater or lesser increases; for example, a steeper slope means a greater increase than a more gradual slope. An ogive, however, is not the ideal graphic for showing comparisons between categories because it simply combines the values in each category and thus indicates an *accumulation*, a growing or lessening total. If you simply wish to keep track of a total and your individual values are periodically combined, an ogive is an appropriate display.

For example, if you saved $300 in both January and April and $100 in each of February, March, May, and June, an ogive would look like this:

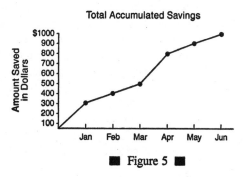

■ Figure 5 ■

An ogive displays a *running total*. Although each individual month's savings could be expressed in a bar chart, you could not easily see the amount of total growth or loss, as you can in an ogive:

■ Figure 6 ■

The choice of graphic display, therefore, depends on what information is important for your purposes: percentages (parts of the whole), running total, comparisons of categories, and so forth.

Frequency Histogram

One of the more commonly used pictorials in statistics is the **frequency histogram,** which in some ways is similar to a bar chart and tells how many items are in each numerical category. For example, suppose that after a garage sale, you want to determine which items were the most popular: the high-priced items, the low-priced items, and so forth. Let's say you sold a total of 32 items for the following prices: $1, $2, $2, $2, $5, $5, $5, $5, $7, $8, $10, $10, $10, $10, $11, $15, $15, $15, $19, $20, $21, $21, $25, $25, $29, $29, $29, $30, $30, $30, $35, $35.

The items sold *ranged* in price from $1 to $35. First, divide this **range** of $1 to $35 into a number of categories, called **class intervals.** Typically, no fewer than 5 or more than 20 class intervals work best for a frequency histogram.

Choose the first class interval to include your lowest (smallest value) data, and make sure that no *overlap* exists so that one piece of data does not fall in two class intervals. For example, you would not have your first class interval be $1-$5 and your second class interval $5-$10 because the four items that sold for $5 would belong in both the first and the second intervals. Instead, use $1-$5 for the first interval and $6-$10 for the second. Class intervals are mutually exclusive.

First, make a table of how your data is distributed. The number of observations that falls into each class interval is called the **class frequency.**

DISTRIBUTION OF ITEMS
SOLD AT GARAGE SALE

Class	Interval	Frequency
1	$1-$5	8
2	$6-$10	6
3	$11-$15	4
4	$16-$20	2
5	$21-$25	4
6	$26-$30	6
7	$31-$35	2

Note that each class interval has the same width. That is, $1-$5 has a width of five dollars, inclusive; $6-$10 has a width of five dollars, inclusive; $11-$15 has a width of five dollars, inclusive, and so forth.

From the data, a frequency histogram would look like this:

■ Figure 7 ■

Unlike in a bar chart, the class intervals are drawn immediately adjacent to each other.

Relative Frequency Histogram

A **relative frequency histogram** uses the same information as above but compares each class interval to the total number of items. For example, the first interval ($1-$5) contains 8 out of the total of 32 items, so the relative frequency of the first class interval is 8/32.

DISTRIBUTION OF ITEMS SOLD AT GARAGE SALE

Class	Interval	Frequency	Relative Frequency
1	$1-$5	8	8/32
2	$6-$10	6	6/32
3	$11-$15	4	4/32
4	$16-$20	2	2/32
5	$21-$25	4	4/32
6	$26-$30	6	6/32
7	$31-$35	2	2/32

The only difference between a frequency histogram and a relative frequency histogram is that the vertical axis uses relative or proportional frequency instead of simple frequency:

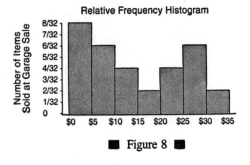

■ Figure 8 ■

Frequency Polygon

Relative frequencies of class intervals can also be shown in a **frequency polygon**. In this chart, the frequency of each class is indicated by points or dots drawn at the midpoints of each class interval. Those points are then connected by straight lines.

Comparing the frequency polygon (Figure 9, below) to the frequency histogram (Figure 7), you will see that the major difference is that points replace the bars:

■ Figure 9 ■

Bar Charts vs. Histograms

Whether to use bar charts or histograms depends on the data. For example, you may have **qualitative** data—numerical information about categories that vary significantly in kind. For instance, gender (male or female), types of automobile owned (sedan, sports car, pickup truck, van, and so forth), and religious affiliations (Christian, Jewish, Moslem, and so forth) are all qualitative data. On the other hand, **quantitative** data can be measured in amounts: age in years, annual salaries, inches of rainfall. Typically, qualitative data are better displayed in bar charts, quantitative data in histograms. As you will see, histograms play an extremely important role in statistics.

Frequency Distribution

Frequency distributions are like frequency polygons (Figure 9); however, instead of straight lines, a frequency distribution uses a smooth curve to connect the points and, similar to a graph, is plotted on two axes: the horizontal axis from left to right (or x axis) indicates the different possible **values** of some **variable** (a phenomenon where observations vary from trial to trial). The vertical axis from bottom to top (or y axis) measures frequency or how many times a particular value occurs.

For example, in Figure 10, the x axis might indicate annual income (the values would be in thousands of dollars); the y axis might indicate frequency (millions of people or percentage of working population). Notice that in Figure 10 the highest percentage of the working population would thus have an annual income in the middle of the dollar values. The lowest percentages would be at the extremes of the values: nearly 0 and extremely high.

Bell-shaped or mound-shaped curve.

■ Figure 10 ■

Notice that this frequency curve displays perfect **symmetry**; that is, one half (the left side) is the mirror image of the other half (the right side). A **bell-shaped** or **mound-shaped curve** is also *normal*, giving it special properties.

Negatively skewed curve.

■ Figure 11 ■

The negatively skewed curve, above, is *skewed to the left*. Its greatest frequency occurs at a value near the right of the graph.

Positively skewed curve.

■ Figure 12 ■

The positively skewed curve, above, is *skewed to the right*. Its greatest frequency occurs at a value near the left of the graph. This distribution is probably a more accurate representation of the annual income of working Americans than is Figure 10.

J-shaped curve.

■ Figure 13 ■

Bimodal curve. Unlike Figure 10, a **bimodal** curve has two high points:

■ Figure 14 ■

Stem-and-Leaf

Another useful pictorial is the **stem-and-leaf**. It is similar to a histogram in that it shows the range of data, where the data are concentrated, if there are any **outliers** (occasional extremely high or extremely low scores), and the general shape of the distribution.

For example, look at the following data—test scores of 17 high school students: 69, 75, 77, 79, 82, 84, 87, 88, 89, 89, 89, 90, 91, 93, 96, 100, 100. The stem-and-leaf takes the first digit of each score as the **stem** and uses the remaining digits as the **leaf**.

As an example, for the score of 69, the 6 is the stem and the 9 is the leaf; for the next three grades (75, 77, and 79), 7 is the stem, and 5, 7, and 9 are the leaves.

Note, too, that along the extreme left side of the chart is a vertical column that keeps a running count or total. (Some stem-and-leafs do not include this running total.) Having a running total enables the reader to locate quickly the **median**. Median is discussed further in Numerical Measures (page 27).

The completed stem-and-leaf for the high school students' test scores looks like this:

STEM-AND-LEAF OF TEST SCORES		
1	6	9
4	7	5 7 9
11	8	2 4 7 8 9 9 9
15	9	0 1 3 6
17	10	0 0

Notice that, like a histogram, each stem determines a class interval and, also like a histogram, the class intervals are all equal. (In this case, each interval width is from a possible low of 0 to a possible high of 9.) All 17 scores are displayed in the stem-and-leaf so that you can see not only the frequencies and the shape of the distribution but also the actual value of every score.

Box Plot (Box-and-Whiskers)

Box plots, sometimes called **box-and-whiskers**, take the stem-and-leaf one step further. A box plot will display a number of values of a distribution of numbers:

1. The median value
2. The lower quartile (Q_1)

3. The upper quartile (Q_3)
4. The interquartile range or *IQR* (distance between the lower and upper quartiles)
5. The symmetry of the distribution
6. The highest and lowest values

Using the following set of values, examine each of the above.

VERBAL SAT SCORES OF 20 STUDENTS

280, 340, 440, 490, 520, 540, 560, 560, 580, 580,
600, 610, 630, 650, 660, 680, 710, 730, 740, 740

The median (the middle value in a set that has been ordered lowest to highest) is the value above which half of the remaining values fall and below which the other half of the remaining values fall. Because there are an even number of scores in our example (20), the median score is the average of the two middle scores (10th and 11th)—580 and 600—or 590.

The **lower quartile** (Q_1 or 25th percentile) is the median of the bottom half. The bottom half of this set consists of the first 10 numbers (ordered from low to high): 280, 340, 440, 490, 520, 540, 560, 560, 580, 580. The median of those 10 is the average of the 5th and 6th scores—520 and 540—or 530. The lower quartile score is 530.

The **upper quartile** (Q_3 or 75th percentile) is the median score of the top half. The top half of this set consists of the last 10 numbers: 600, 610, 630, 650, 660, 680, 710, 730, 740, 740. The median of these 10 is again the average of the 5th and 6th scores—in this case, 660 and 680—or 670. So 670 is the upper quartile score for this set of 20 numbers.

A box plot can now be constructed as follows: the left side of the box indicates the lower quartile; the right side of the box indicates the upper quartile; and the line inside the box indicates the median. A horizontal line is then drawn from the lowest value of the distribution through the box to the highest value of the distribution. (This horizontal line is the "whiskers.")

Using the Verbal SAT scores above, a box plot would look like this:

■ Figure 15 ■

Without reading the actual values, you can see by looking at the box plot (Figure 15) that the scores range from a low of 280 to a high of 740; that the lower quartile (Q_1) is at 530; the median is at 590; and the upper quartile (Q_3) is at 670. Because the median is slightly nearer the lower quartile than the upper quartile and the interquartile range is situated far to the right of the range of values, the distribution departs from symmetry.

Scatterplot

Sometimes you wish to display information about the relationship involving two different phenomena. For example, suppose you collected data about the number of days that law school candidates studied for a state bar examination and their resulting scores on the exam. The data from eight candidates is as follows:

Candidate #	1	2	3	4	5	6	7	8
Days studied	7	9	5	1	8	4	3	6
Score earned	23	25	14	5	22	15	11	17

One dot would then be plotted for each examinee, giving a total of only 8 dots, yet displaying 16 pieces of numerical information. For example, candidate #1 studied for 7 days and received a score of 23. Candidate #1's dot would be plotted at a vertical of 23 and a horizontal of 7:

Relationship Between
Days Studied and Bar Exam Scores

■ Figure 16 ■

A completed scatterplot would look like this:

Figure 17

There is a strong **positive relationship** between the number of days studied and the score on the bar exam; that is, the data displayed indicates that an increase in days studied for the exam correlates with an increase in score achieved. A **negative relationship** would be indicated if the dots suggest a line going down from left to right, meaning that as one variable increases, the other decreases. And no relationship would be indicated if the scatterplot dots suggest a completely horizontal line, a completely vertical line, or no line at all.

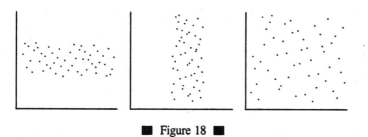

Figure 18

These relationships are discussed in more detail in Bivariate Relationships (page 111).

Authors' Note: The graphic displays presented in this section are among the most commonly encountered in statistical analysis.

This section discusses how to graph numerical information—to transform it into a picture—to get a better sense of the data. Another way to get a sense of data is to use numerical measures, certain numbers that give special insight into your values. Two types of numerical measures are important in statistics: **measures of central tendency** and **measures of variation**. Each of these individual measures can provide information about the entire set of data.

Measures of Central Tendency

Measures of central tendency are numbers which tend to cluster around the "middle" of a set of values. Three such middle numbers are the **mean**, the **median**, and the **mode**.

For example, suppose your earnings for the past week were

Monday	$350
Tuesday	$150
Wednesday	$100
Thursday	$350
Friday	$50

Mean. You could express your daily earnings of the above set in a number of ways. One way is to use the average, or **mean,** of each day. The arithmetic mean is the sum of the measures in the set divided by the number of measures in the set. Totaling all the measures and dividing by the number of measures, you get $1,000 divided by 5 = $200.

Median. Another measure of central tendency is the **median,** which is defined as the middle value when the numbers are arranged in increasing or decreasing order. When you order the daily earnings above, you get: $50, $100, $150, $350, $350. The middle value is $150, and therefore, $150 is the median.

If there is an even number of items in a set, the median is the average of the two middle values. For example, if we had four values—4, 10, 12, 26—the median would be the average of the two middle values, 10 and 12; thus, 11 is the median. The median may sometimes be a better indicator of central tendency than the mean, especially when there are **outliers,** or extreme values.

For example, given the following four annual salaries of a corporation:

CEO	$1,000,000
Manager	50,000
Administrative	30,000
Clerical	20,000

The mean of these four salaries is $275,000. The median is the average of the middle two salaries, or $40,000. In this instance, the median appears to be a better indicator of central tendency because the CEO's salary is an extreme outlier causing the mean to lie far from the other three salaries.

Mode. Another indicator of central tendency is the **mode,** or the value that occurs most often in a set of numbers. In the set of weekly earnings on page 27, the mode would be $350 because it appears twice and the other values appear only once.

Notation and formulae. The mean of a sample is typically denoted as $\bar{x}$ (read as x bar). The mean of a population is typically denoted as μ (read as mew.) The sum (or total) of measures is typically denoted with a Σ. The formula for a sample mean is

$$\bar{x} = \frac{\Sigma x}{n} = \frac{x_1 + x_2 + \ldots x_n}{n}$$

Mean for grouped data. Occasionally you may have data that do not consist of actual values but rather **grouped measures**. For example, you may know that in a certain working population, 32% earn between $25,000 and $29,999, 40% earn between $30,000 and $34,999, 27% earn between $35,000 and $39,999, and the remaining 1% earn between $80,000 and $85,000. This type of information is similar to that presented in a frequency table. (See Graphic Displays, page 9.) Although you do not have precise individual measures, you can, nevertheless, compute measures for **grouped data,** data presented in a frequency table.

The formula for a sample mean for grouped data is

$$\bar{x} = \frac{\Sigma fx}{n}$$

where x denotes the midpoint of the interval; fx denotes the sum of the measurements of the interval.

For example, if 8 is the midpoint of a class interval and there are 10 measurements in the interval, $fx = 10(8) = 80$, the sum of the 10 measurements in the interval.

Σfx denotes the sum of all the measurements in *all* class intervals. Dividing that sum by the number of measurements yields the sample mean for grouped data.

For example, consider the following information from Graphic Displays:

Class Interval	Frequency (f)	Midpoint (x)	fx
$1–$5	8	3	24
$6–$10	6	8	48
$11–$15	4	13	52
$16–$20	2	18	36
$21–$25	4	23	112
$26–$30	6	28	168
$31–$35	2	33	66
	$n = 32$		$\Sigma fx = 506$

Substituting into the formula:

$$\bar{x} = \frac{\Sigma fx}{n} = \frac{506}{32} = 15.81$$

therefore, the average price of items sold was about $15.81.

Median for grouped data. The median for grouped data may not necessarily be computed precisely because the actual values of the measurements may not be known. In that case, you can find the particular interval that contains the median, and approximate the median.

Using the table above, you can see that there is a total of 32 measures. The median is between the 16th and 17th measure, and therefore the median falls within the $11-$15 interval. The formula for the best approximation of the median for grouped data is

$$\text{median} = L + \frac{w}{f_{med}}(.5n + \Sigma f_b)$$

where L = lower class limit of the interval that contains the median

n = total number of measurements

w = interval width

f_{med} = frequency of the class containing the median

Σf_b = sum of the frequencies for all classes before the median class

Consider the information:

PRICES OF ITEMS SOLD AT GARAGE SALE	
Class Interval	Frequency (f)
$1–$5.99	8
$6–$10.99	6
$11–$15.99	4
$16–$20.99	2
$21–$25.99	4
$26–$30.99	6
$31–$35	2
	n = 32

As we already know, the median is located in class interval $11 -$15. So $L = 11$, $n = 32$, $w = 4.99$, $f_{med} = 4.99$, $\Sigma f_b = 14$.

Substituting into the formula:

$$\text{median} = L + \frac{w}{f_{med}}\ (.5n - \Sigma f_b)$$

$$= 11 + \frac{4.99}{4.99}\ (.5(32) - 14)$$

$$= 11 + \frac{4.99}{4.99}\ (16 - 14)$$

$$= 11 + \frac{4.99}{4.99}\ (2)$$

$$= 11 + 2 = 13$$

Symmetric distribution. In a distribution displaying perfect symmetry, the mean, the median, and the mode are all at the same point:

Mean
Median
Mode

■ Figure 19 ■

Skewed curves. As you have seen, an outlier can significantly alter the mean of a series of numbers, whereas the median will remain at the center of the series. In such a case, the resulting curve drawn from the values will appear to be **skewed**, tail off rapidly to the left or right. (See Graphic Displays, page 19.) In the case of negatively skewed or positively skewed curves, the median remains in the center of these three measures.

Negatively skewed:

■ Figure 20 ■

Positively skewed:

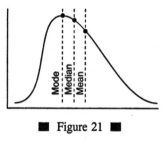

■ Figure 21 ■

Measures of Variability

Measures of central tendency locate only the center of a distribution of measures. Other measures are often needed to describe data.

For example, consider these two sets of numbers:

Earnings of Employee A/Day	Earnings of Employee B/Day
$200	$200
$210	$20
$190	$400
$201	$0
$199	$390
$195	$10
$205	$200
$200	$380

The mean, the median, and the mode of each employee's daily earnings all equal $200. Yet, there is significant difference between the two sets of numbers. For example, the daily earnings of Employee A are much more consistent than those of Employee B, which show great variation. This example illustrates the need for **measures of variation** or spread.

Range. The most elementary measure of variation is **range**. Range is defined as the difference between the largest and smallest values. Note that the range for Employee A is $210 - $190 = $20; the range for Employee B is $400 - $0 = $400.

Deviation and variance. The deviation is defined as the distance of the measurements away from the mean. In the table above, Employee A's earnings have considerably less deviation than do Employee B's. The **variance** is defined as the sum of the squared deviations of n measurements from their mean divided by $(n - 1)$.

So, from the table of employee earnings, the mean for Employee A is $200, and the deviations from the mean are:

$$0, +10, -10, +1, -1, -5, +5, 0$$

The squared deviations from the mean are, therefore:

$$0, 100, 100, 1, 1, 25, 25, 0$$

The sum of these squared deviations from the mean equals 252. Divided by $(n - 1)$, or $8 - 1$, yields 252/7, which equals 36. The variance $= 36$.

For Employee B, the mean is also $200, and the deviations from the mean are:

$$0, -180, +200, -200, +190, -190, 0, +180$$

The squared deviations are, therefore:

$$0; 32,400; 40,000; 40,000, 36,100; 36,100; 0; 32,400$$

The sum of these squared deviations equals 217,000. Divided by $(n - 1)$ yields 217,000/7, which equals 31,000.

Although they earned the same totals, there is significant difference in variance between the daily earnings of the two employees.

Standard deviation. The **standard deviation** is defined as the positive square root of the variance; thus, the standard deviation of Employee A's daily earnings is the positive square root of 36, or 6. The standard deviation of Employee B's daily earnings is the positive square root of 31,000, or about 176.

Notation.

s^2 denotes the variance of a sample.

σ^2 denotes the variance of a population.

s denotes the standard deviation of a sample.

σ denotes the standard deviation of a population.

Empirical rule: the normal curve. The practical significance of the standard deviation is that with mound-shaped (bell-shaped) distributions, the following rules apply;

1. The interval from one standard deviation below the mean to one standard deviation above the mean contains approximately 68% of the measurements.
2. The interval from two standard deviations below the mean to two standard deviations above the mean contains approximately 95% of the measurements.
3. The interval from three standard deviations below the mean to three standard deviations above the mean contains approximately all of the measurements.

These mound-shaped curves are usually called **normal distributions** or **normal curves.**

■ Figure 22 ■

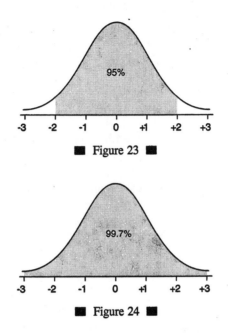

■ Figure 23 ■

■ Figure 24 ■

Shortcut formulae. A shortcut method of calculating variance and standard deviation requires two quantities: sum of the values and sum of the squares of the values.

Σx = sum of the measures

Σx^2 = sum of the squares of the measures

For example, using these six measures: 3, 9, 1, 2, 5, and 4:

$$\Sigma x = 3 + 9 + 1 + 2 + 5 + 4 = 24$$
$$\Sigma x^2 = 3^2 + 9^2 + 1^2 + 2^2 + 5^2 + 4^2$$
$$= 9 + 81 + 1 + 4 + 25 + 16 = 136$$

The quantities are then substituted into the shortcut formula to find $\Sigma(x - \bar{x})^2$.

$$\Sigma(x - \bar{x})^2 = \Sigma x^2 - \frac{(\Sigma x)^2}{n}$$

$$= 136 - \frac{(24)^2}{6}$$

$$= 136 - \frac{576}{6} = 40$$

The variance and standard deviation are now found as before:

$$s^2 = \frac{\Sigma(x - \bar{x})^2}{n - 1} = \frac{40}{5} = 8$$

$$s = \sqrt{s^2} = \sqrt{8} = 2.828$$

Guessing the standard deviation. To avoid tedious calculations, you can quickly obtain an estimate of the standard deviation by using the formula:

$$s = \frac{r}{4}$$

which reads "the standard deviation is approximately equal to the range divided by 4."

Consider the 6 measures listed above. The range was 9 – 1, or 8. Dividing by 4 gives an approximate standard deviation of about 2.

Percentile. The Nth **percentile** is defined as the value such that N% of the values lie below it. So, for example, a score 5% from the top score would be the 95th percentile because it is above 95% of the other scores.

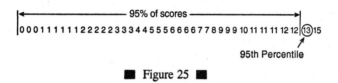

Figure 25

Quartiles and interquartile range. The **lower quartile** (Q_1) is defined as the 25th percentile; thus, 75% of the measures are above the lower quartile. The **middle quartile** (Q_2) is defined as the 50th percentile, which is, in fact, the median of all the measures. The **upper quartile** (Q_3) is defined as the 75th percentile; thus, only 25% of the measures are above the upper quartile.

The **interquartile range** *(IQR)* is the set of measures lying between the upper and lower quartiles, or the middle 50% of the measures.

Figure 26

Measurement Scales

Different measurement scales allow for different levels of exactness, depending upon the characteristics of the variables being measured. The four types of scales available in statistical analysis are

1. **Nominal:** A scale that measures data by name only. For example, religious affiliation (measured as Jewish, Christian, Buddhist, and so forth), political affiliation (measured as Democratic, Republican, Libertarian, and so forth), or style of automobile (measured as sedan, sports car, station wagon, van, and so forth).

2. **Ordinal:** Measures by rank order only. Other than rough order, no precise measurement is possible. For example, medical condition (measured as satisfactory, fair, poor, guarded, serious, and critical); social-economic status (measured as lower class, lower-middle class, middle class, upper-middle class, upper class); or military officer rank (measured as lieutenant, captain, major, lieutenant colonel, colonel, general). Such rankings are not absolute but rather "relative" to each other: major is higher than captain, but we cannot measure the exact difference in numerical terms. Is the difference between major and captain equal to the difference between colonel and general? You cannot say.

3. **Interval:** Measures by using equal intervals. Here you can compare differences between pairs of values. The Fahrenheit temperature scale, measured in degrees, is an interval scale, as is the centigrade scale. The temperature difference between 50 and 60 degrees centigrade (10 degrees) equals the temperature difference between 80 and 90 degrees centigrade (10 degrees). Note that the 0 in each of these scales is arbitrarily placed, which makes the interval scale different from ratio.

4. **Ratio:** Similar to an interval scale, a ratio scale includes a 0 measurement that signifies the point at which the characteristic being measured vanishes (absolute 0). For example, income (measured in dollars, with 0 equal to no income at all), years of formal education, items sold, and so forth, are all ratio scales.

Probability theory plays an important role in statistics. This section explores three areas of probability: simple events, combinations of events, and the binomial distribution.

Classic Theory

The classic theory of probability underlies much of probability in statistics. Briefly, it states that the chance of a particular outcome occurring is determined by the ratio of the number of favorable outcomes (or "successes") to the total number of outcomes. Expressed as a formula,

$$P(A) = \frac{\text{number of favorable outcomes}}{\text{total number of possible outcomes}}$$

For example, the probability of randomly drawing an ace from a well-shuffled deck of cards is equal to the ratio 4/52. Four is the number of favorable outcomes (the number of aces in the deck), and 52 is the number of total outcomes (the number of cards in the deck). The probability of randomly selecting an ace in one draw from a deck of cards is therefore 4/52, or .077. In statistical analysis, probability is usually expressed as a decimal and ranges from a low of 0 (no chance) to a high of 1.0 (certainty).

The classic theory assumes that all outcomes have equal likelihood of occurring. In the example just cited, each card must have an equal chance of being chosen—no card is larger than any other or in any way more likely to be chosen than any other card.

The classic theory pertains only to outcomes that are **mutually exclusive** (or **disjoint**), which means that those outcomes may not occur at the same time. For example, one coin flip can result in a head *or* a tail, but one coin flip cannot result in a head *and* a tail. So the outcome of a head and the outcome of a tail are said to be "mutually exclusive" in one coin flip, as is the outcome of an ace *and* a king as the outcome of one card being drawn.

Relative Frequency Theory

The **relative frequency theory of probability** holds that if an experiment is repeated an extremely large number of times, and a particular outcome occurs a percentage of the time, then that particular percentage is close to the probability of that outcome.

For example, if a machine produces 10,000 widgets one at a time, and 1,000 of those widgets are faulty, the probability of that machine producing a faulty widget is approximately 1,000 out of 10,000, or .10.

Probability of Simple Events

What is the probability of simultaneously flipping three coins—a penny, a nickel, and a dime—and having all three land heads?

Using the classic theory, determine the ratio of number of favorable outcomes to the number of total outcomes. Listed below are all possible outcomes:

Outcome #	Penny	Nickel	Dime
1	H	H	H
2	H	H	T
3	H	T	H
4	H	T	T
5	T	H	H
6	T	H	T
7	T	T	H
8	T	T	T

There are 8 different outcomes, only one of which is favorable (outcome #1: all three coins landing heads); therefore, the probability of three coins landing heads is 1/8, or .125.

What is the probability of exactly two of the three coins landing heads? Again, there are the 8 total outcomes, but in this case only 3 favorable outcomes (outcomes #2, #3, and #5); thus, the probability of exactly two of three coins landing heads is 3/8 or .375.

Independent Events

Each of the three coins being flipped above is what is known as an **independent event**. Independent events are defined as outcomes that are not affected by other outcomes. In other words, the flip of the penny does not affect the flip of the nickel, and vice versa.

Dependent Events

Dependent events, on the other hand, are outcomes that are affected by other outcomes. For example, what is the probability of randomly drawing an ace from a deck of cards and then drawing an ace again from the same deck of cards, without returning the first drawn card back to the deck?

For the first draw, the probability of a favorable outcome is 4/52, as explained above; however, once that first card has been drawn, the total number of outcomes is no longer 52, but now 51 because a card has been removed from the deck. And if that first card drawn resulted in a favorable outcome (an ace), there would now be only 3 aces in the deck. Or the number of favorable outcomes would remain at 4 if that first card drawn were not an ace. So the second draw is a dependent event because its probability changes depending upon what happens on the first draw.

If, however, you replace that drawn card back into the deck *and* shuffle well again before the second draw, then the probability for a favorable outcome for each draw will now be equal (4/52) and these events will be independent.

Probability of Joint Occurrences

Another way to compute the probability of all three flipped coins landing heads is as a series of three different events: first flip the penny, then flip the nickel, then flip the dime. Will the probability of landing three heads still be .125?

Multiplication rule. To compute the probability of two or more independent events *all* occurring—**joint occurrence**—multiply their probabilities.

For example, the probability of the penny landing heads is 1/2, or .5; the probability of the nickel next landing heads is 1/2, or .5; and the probability of the dime landing heads is 1/2, or .5; thus, note that

$$.5 \times .5 \times .5 = .125$$

which is what you determined with the classic theory by assessing the ratio of number of favorable outcomes to number of total outcomes. The notation for joint occurrence is $P(AB) = P(A) \times P(B)$ and reads: The probability of A and B both happening is equal to the probability of A times the probability of B.

Using the multiplication rule, you can also determine the probability of drawing two aces in a row from a deck of cards. The only way to draw two aces in a row from a deck of cards is for both draws to be favorable. For the first draw, the probability of a favorable outcome is 4/52. But because the first draw is favorable, only 3 aces are left among 51 cards. So the probability of a favorable outcome on the second draw is 3/51. For both events to happen, you simply multiply those two probabilities together:

$$\frac{4}{52} \times \frac{3}{51} = \frac{12}{2652} = .0045$$

Note that these probabilities are not independent. If, however, you had decided to return the initial card drawn back to the deck before the second draw, then the probability of drawing an ace on each draw is 4/52, as these events are now independent. Drawing an ace twice in a row, with the odds being 4/52 both times, gives:

$$\frac{4}{52} \times \frac{4}{52} = \frac{16}{2704} = .0059$$

In either case, you use the multiplication rule because you are computing probability for favorable outcomes in *all* events.

Addition rule. Given mutually exclusive events, finding the probability of *at least one* of them occurring is accomplished by adding their probabilities.

For example, what is the probability of one coin flip resulting in at least one head or at least one tail?

The probability of one coin flip landing heads is .5, and the probability of one coin flip landing tails is .5. Are these two outcomes mutually exclusive in one coin flip? Yes, they are. You cannot have a coin land both heads and tails in one coin flip; therefore, you can determine the probability of at least one head or one tail resulting from one flip by adding the two probabilities:

$$.5 + .5 = 1.0 \text{ (or certainty)}$$

As another example, what is the probability of at least one spade or one club being randomly chosen in one draw from a deck of cards?

The probability of drawing a spade in one draw is 13/52; the probability of drawing a club in one draw is 13/52. These two outcomes are mutually exclusive in one draw because you cannot draw both a spade and a club in one draw; therefore, you can use the addition rule to determine the probability of drawing at least one spade or one club in one draw:

$$\frac{13}{52} + \frac{13}{52} = \frac{26}{52} = .50$$

Non-mutually Exclusive Outcomes

For the addition rule to apply, the events must be mutually exclusive. Now consider the following example.

What is the probability of the outcome of at least one head in *two* coin flips?

Should you add the two probabilities as in the examples above? In the first example, you added the probability of getting a head and the probability of getting a tail because those two events were mutually exclusive in one flip. In the second example, the probability of getting a spade was added to the probability of getting a club because those two outcomes were mutually exclusive in one draw. Now when you have two flips, should you add the probability of getting a head on the first flip to the probability of getting a head on the second flip? Are these two events mutually exclusive?

Of course, they are not mutually exclusive. You *can* get an outcome of a head on one flip *and* a head on the second flip. So, because they are not mutually exclusive, you cannot use the addition rule. (If you did use the addition rule, you would get

$$\frac{1}{2} + \frac{1}{2} = \frac{2}{2} = 1.0$$

or certainty, which is absurd. There is no certainty of getting at least one head on two flips. Try it several times, and see that there is a possibility of getting two tails and no heads.)

"Double Counting"

By using the addition rule in a situation that is not mutually exclusive, you are in fact "double counting." One way of realizing this double counting is to use the classic theory of probability: list all the different outcomes when flipping a coin twice, and assess the ratio of favorable outcomes to total outcomes:

First Flip	with	Second Flip
head	+	head
head	+	tail
tail	+	head
tail	+	tail

There are 4 total outcomes. Three of the outcomes have at least one head; therefore, the probability of throwing at least one head in two flips is 3/4 or .75, not 1.0. But if you had used the addition rule, you would have added the two heads from the first flip to the two heads from the second flip and gotten 4 heads in 4 flips, 4/4 = 1.0! But the two heads in that first pair constitute only one outcome, and so by counting both heads for that outcome, you are "double counting" because this is the joint-occurrence outcome that is not mutually exclusive.

To use the addition rule in a non-mutually exclusive situation, you must subtract any events that double count. In this case:

$$\frac{1}{2} + \frac{1}{2} - \frac{1}{4} = \frac{3}{4} = .75$$

The notation, therefore, for at least one favorable occurrence in two events is

$$P(A + B) = P(A) + P(B) - P(AB)$$

This rule is read: The probability of at least one of the events A or B equals the probability of A plus the probability of B minus the probability of their joint occurrence. (Note that if they are mutually exclusive, then $P(AB)$—the joint occurrence—equals 0, and you simply add the two probabilities.)

For example, what is the probability of drawing either a spade or an ace from a deck of cards?

The probability of drawing a spade is 13/52, the probability of drawing an ace is 4/52. But the probability of their joint occurrence (an ace of spades) is 1/52. Thus,

$$P(A + B) = P(A) + P(B) - P(AB)$$

$$= \frac{13}{52} + \frac{4}{52} - \frac{1}{52}$$

$$= \frac{16}{52} = \frac{4}{13} = .3077$$

Conditional Probability

Sometimes you have more information than simply total outcomes and favorable outcomes and, hence, are able to make more informed judgments regarding probabilities. For example, suppose you know the following information.

In a particular village, there are 60 women and 40 men. Twenty of those women are 70 years of age or older; 5 of the men are 70 years of age or older.

	70+ yrs	69 or less	Totals
Women	20	40	60
Men	5	35	40
Totals	25	75	100

What is the probability that a person selected at random in that town will be a woman? Because women constitute 60% of the total population, the probability is .60.

What is the probability that a person 70+ years of age selected at random will be a woman? This question is different because the probability of A (being a woman) *given B* (the person in question is 70+ years of age) is now conditional upon B (being 70+ years of age). Because women number 20 out of the 25 people in the 70+ years-old group, the probability of this latter question is 20/25, or .80.

Conditional probability is found using the formula:

$$P(A|B) = \frac{AB}{B} = \frac{P(AB)}{P(B)}$$

which is read: The probability of A given B equals the proportion of the total of A and B to the total of B. The vertical bar in the expression A|B is read *given that* or *given*.

Probability Distributions

A **probability distribution** is a pictorial display of the probability—$P(x)$—for any value of x. Consider the number of possible outcomes of two coins being flipped:

POSSIBLE OUTCOMES OF
TWO FLIPPED COINS

H + H	=	2 heads
H + T	=	1 head
T + H	=	1 head
T + T	=	0 heads

PROBABILITY DISTRIBUTION:
NUMBER OF HEADS

x	$P(x)$
0	1/4 or .25
1	1/2 or .50
2	1/4 or .25

■ Figure 27 ■

Discrete vs. continuous variables. The number of heads resulting from coin flips can be counted only in integers (whole numbers). The number of aces drawn from a deck can be counted only in integers. These "countable" numbers are known as **discrete variables**: 0, 1, 2, 3, 4, and so forth. Nothing in between two variables is possible. For example, 2.6 is not possible.

Qualities such as height, weight, temperature, distance, and so forth, however, can be measured using fractions or decimals as well: 34.27, 102.26, and so forth. These are known as **continuous variables.**

Total of probabilities of discrete variables. As you have noted, the probability of a discrete variable lies somewhere between 0 and 1, inclusive. For example, the probability of tossing one head in two coin flips is, as above, .50. The probability of tossing two heads in two coin flips is .25, and the probability of tossing no heads in two coin flips is .25. The sum (total) of probabilities for all values of x always equals 1. For example, note that in the two tables above, adding the probabilities for all the possible outcomes yields a sum of 1.

The Binomial

A discrete variable that can result in only one of two outcomes is called **binomial**. For example, a coin flip is a binomial variable; but drawing a card from a standard deck of 52 is not. Whether a drug is either successful or unsuccessful in producing results is a binomial variable, as is whether a machine produces perfect or imperfect widgets.

Binomial experiments. Binomial experiments require the following elements:

1. The experiment consists of a number of identical events (n).
2. Each event has only one of two mutually exclusive outcomes. (These outcomes are called successes and failures.)
3. The probability of a "success" outcome is equal to some percentage, which is identified as a **proportion, π**.
4. This proportion, π, remains constant throughout all events and is defined as the ratio of number of successes to number of trials.
5. The events are independent.
6. Given all of the above, the **binomial** formula can be applied (x = number of favorable outcomes; n = number of events):

$$P(x) = \frac{n!}{x! \ (n - x)!} \ \pi^x (1 - \pi)^{n - x}$$

$$n! = n(n - 1)(n - 2) \ldots (3)(2)(1)$$

Consider this example: A coin is flipped 10 times. What is the probability of getting exactly 5 heads? Using the binomial formula, where n (the number of events) is given as 10; x (the number of favorable outcomes) is given as 5; and the probability of landing a head in one flip is .5:

$$P(x) = \frac{n!}{x!(n - x)!} \ \pi^x(1 - \pi)^{n - x}$$

$$= \frac{10!}{5!(5)!}(.5^5)(1 - .5)^5$$

$$= \frac{10 \cdot 9 \cdot 8 \cdot 7 \cdot 6 \cdot 5 \cdot 4 \cdot 3 \cdot 2 \cdot 1}{5 \cdot 4 \cdot 3 \cdot 2 \cdot 1(5 \cdot 4 \cdot 3 \cdot 2 \cdot 1)}(.0313)(.0313)$$

$$= 252(.0313)(.0313)$$

$$= .246$$

So, the probability of getting exactly 5 heads in 10 flips is .246, or approximately 25%.

Binomial table. Because probabilities of binomial variables are so common in statistics, tables are used to alleviate having to continually use the formula. Refer to Table 1 on page 138, and you will find that given $n = 10$, $x = 5$, and $\pi = .5$, the probability is .2461.

Mean and standard deviation. The **mean** of the **binomial probability distribution** is determined by the formula:

$$\mu = n\pi$$

where π is the proportion of favorable outcomes and n is the number of events.

The **standard deviation of the binomial probability distribution** is determined by the formula:

$$\sigma = \sqrt{n\pi(1 - \pi)}$$

For example, what is the mean and standard deviation for a binomial probability distribution for 10 coin flips of a fair coin?

Because the proportion of favorable outcomes of a fair coin falling heads (or tails) is $\pi = .5$, simply substitute into the formulas:

$$\mu = n\pi = 10(.5) = 5$$
$$\sigma = \sqrt{n\pi(1 - \pi)} = \sqrt{(10)(.5)(.5)} = \sqrt{2.5} = 1.58$$

The probability distribution for the number of favorable outcomes is

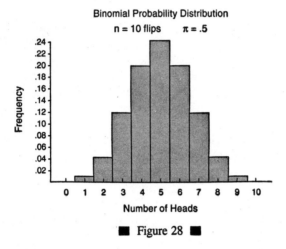

■ Figure 28 ■

Note that this distribution appears to display symmetry. Binomial probability distributions and use of the normal approximation to replace the probability histogram are discussed in Sampling (page 55).

As discussed in the Overview (page 3), the field of inferential statistics enables you to make educated guesses about the numerical characteristics of large groups. The logic of sampling gives you a way to test conclusions about such groups using only a small portion of its members.

Populations, Samples, Parameters, and Statistics

A **population** is a group of phenomena that have something in common. The term often refers to a group of people:

- All registered voters in Crawford County
- All members of the International Machinists Union
- All Americans who played golf at least once in the past year

But populations can refer to things as well as people:

- All widgets produced last Tuesday by the Acme Widget Company
- All daily maximum temperatures in July for major U.S. cities
- All basal ganglia cells from a particular rhesus monkey

Often, researchers want to know things about populations but don't have data for every person or thing in the population. If a company's customer service division wanted to learn whether its customers were satisfied, it would not be practical (or perhaps even possible) to contact every individual who purchased one of their products. Instead, they might select a **sample** of the population. A sample is a smaller group of members of a population selected to represent the population. In order to use statistics to learn things about the population, the sample must be **random**. A random sample is one in which every member of a population has an equal chance to be selected.

A **parameter** is a characteristic of a population. A **statistic** is a characteristic of a sample. Inferential statistics enables you to make an educated guess about a population parameter based on a statistic computed from a sample randomly drawn from that population.

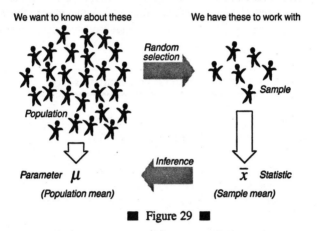

We want to know about these   We have these to work with

Random selection

Population

Inference

Parameter μ

(Population mean)

Sample

$\bar{x}$ Statistic

(Sample mean)

■ Figure 29 ■

For example, say you want to know the mean income of the subscribers to a particular magazine—a parameter of a population. You draw a random sample of 100 subscribers and determine that their mean income $\bar{x}$ is $27,500 (a statistic). You conclude that the population mean income μ is likely to be close to $27,500 as well. This example is one of statistical inference.

Different symbols are used to denote statistics and parameters:

	Sample Statistic	Population Parameter
Mean	$\bar{x}$	μ
Standard deviation	s, sd	σ
Variance	s^2	σ^2

Sampling Distributions

Continuing with the above example, suppose that ten different samples of 100 people were drawn from the population, instead of just one. The income means of these ten samples would not be expected to be exactly the same, because of **sampling variability**. Sampling variability is the tendency of the same statistic computed from a number of random samples drawn from the same population to differ.

Suppose that the first sample of 100 magazine subscribers was "returned" to the population (made available to be selected again), another sample of 100 subscribers selected at random, and the mean income of the new sample computed. If this process were repeated ten times, it might yield the following sample means:

27,500	27,192	28,736	26,454	28,527
28,407	27,592	27,684	28,827	27,809

These ten values are part of a **sampling distribution**. The sampling distribution of a statistic—in this case, of a mean—is the distribution obtained by computing the statistic for a large number of samples drawn from the same population.

You can compute the sample mean of this sampling distribution by summing the ten sample means and dividing by ten, which gives a distribution mean of 27,872. Suppose that the mean income of the entire population of subscribers to the magazine is $28,000. (You usually do not know what it is.) You can see in Figure 30 that the first sample mean—$27,500—was not a bad estimate of the population mean and that the mean of the distribution of ten sample means—$27,872— was even better.

Distribution of population
Distribution of
first sample of 100
Distribution of
means from 10 samples
(part of a sampling
distribution)

28,000 (Actual population mean)
27,872 (Mean of distribution of 10 means)
27,500 (Mean of first sample)

■ Figure 30 ■

The more sample means included in the sampling distribution, the more accurate the mean of the sampling distribution becomes as an estimate of the population mean.

Random and Systematic Error

Two potential sources of error occur in statistical estimation—two reasons a statistic might misrepresent a parameter. **Random error** occurs as a result of sampling variability. The ten sample means above differed from the true population mean because of random error. Some were below the true value; some above it. Similarly, the mean of the distribution of ten sample means was slightly lower than the true population mean. If ten more samples of 100 subscribers were drawn, the mean of that distribution—that is, the mean of those means—might be higher than the population mean.

Systematic error or **bias** refers to the tendency to consistently underestimate or overestimate a true value. Suppose that your list of magazine subscribers was obtained through a database of information about air travelers. The samples that you would draw from such a list would likely overestimate the population mean of all subscribers' income because lower-income subscribers are less likely to travel by air and many of them would be unavailable to be selected for the samples. This example would be one of bias.

Figure 31

Central Limit Theorem

If the population of all subscribers to the magazine were normal, you would expect its sampling distribution of means to be normal as well. But what if the population were non-normal? The **Central Limit Theorem** states that even if a population distribution is strongly non-normal, its sampling distribution of means will be approximately normal for large sample sizes (over 30). The Central Limit Theorem makes it possible to use probabilities associated with the normal curve to answer questions about the means of sufficiently large samples.

According to the Central Limit Theorem, the mean of a sampling distribution of means is an unbiased estimator of the population mean.

$$\mu_{\bar{x}} = \mu$$

Similarly, the standard deviation of a sampling distribution of means is

$$\sigma_{\bar{x}} = \frac{\sigma}{\sqrt{n}}$$

Note that the larger the sample, the less variable the sample mean. The mean of many observations is less variable than the mean of few. The standard deviation of a sampling distribution of means is often called the **standard error** of the mean. Every statistic has a standard error, which is a measure of the statistic's random variability.

For example, if the population mean of number of fish caught per trip to a particular fishing hole is 3.2 and the population standard deviation is 1.8, what are the population mean and standard deviation of 40 trips?

$$\mu_{\bar{x}} = 3.2$$

$$\sigma_{\bar{x}} = \frac{1.8}{\sqrt{40}} = .285$$

Properties of the Normal Curve

Known characteristics of the normal curve make it possible to estimate the probability of occurrence of any value of a normally distributed variable. Suppose that the total area under the curve is defined to be 1. You can multiply that number by 100 and say there is a 100% chance that any value you can name will be somewhere in the distribution. (Remember, the distribution extends to infinity in both directions.) Similarly, because half of the area of the curve is below the mean and half is above it, you can say that there's a 50% chance that a randomly chosen value will be above the mean and the same chance that it will be below it.

It makes sense that the area under the normal curve is equivalent to the probability of randomly drawing a value in that range. The area is greatest in the middle, where the "hump" is, and thins out toward the tails. That's consistent with the fact that there are more values close to the mean in a normal distribution than far from it.

When the area of the standard normal curve is divided into sections by standard deviations above and below the mean, the area in each section is a known quantity (Figure 32). As explained above, the area in each section is the same as the probability of randomly drawing a value in that range.

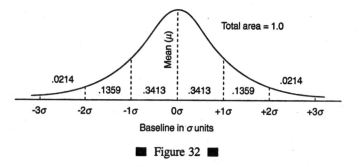

■ Figure 32 ■

For example, .3413 of the curve falls between the mean and one standard deviation above the mean, which means that about 34% of all the values of a normally distributed variable are between the mean and one standard deviation above it. It also means that there is a .3413 chance that a value drawn at random from the distribution will lie between these two points.

Sections of the curve above and below the mean may be added together to find the probability of obtaining a value within (plus or minus) a given number of standard deviations of the mean (Figure 33). For example, the amount of curve area between one standard deviation above the mean and one standard deviation below is .3413 + .3413 = .6826, which means that exactly 68.26% of the values lie in that range. Similarly, about 95% of the values lie within two standard deviations of the mean, and 99.7% of the values lie within three standard deviations.

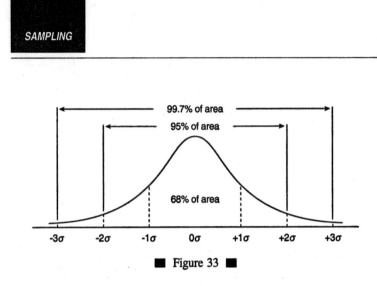

■ Figure 33 ■

In order to use the area of the normal curve to determine the probability of occurrence of a given value, the value must first be **standardized**, or converted to a z-score. To convert a value to a z-score is to express it in terms of how many standard deviations it is above or below the mean. After the z-score is obtained, its corresponding probability may be looked up in a table. The formula to compute a z-score is

$$z = \frac{x - \mu}{\sigma}$$

where x is the value to be converted, μ is the population mean, and σ is the population standard deviation.

For example, a normal distribution of retail store purchases has a mean of \$14.31 and a standard deviation of 6.40. What percentage of purchases were under \$10? First, compute the z-score:

$$z = \frac{10 - 14.31}{6.40} = -.67$$

The next step is to look up the z-score in the table of standard normal probabilities (Table 2, page 150). The standard normal table lists the probabilities (curve areas) associated with given z-scores.

Table 2 gives the area of the curve below z or, in other words, the probability of obtaining a value of z or lower. Not all standard normal tables use the same format, however. Some list only positive z-scores and give the area of the curve between the mean and z. Such a table is slightly more difficult to use, but the fact that the normal curve is symmetric makes it possible to use it to determine the probability associated with any z-score, and vice versa.

To use Table 2, first look up the z-score in the left column, which lists z to the first decimal place. Then look along the top row for the second decimal place. The intersection of the row and column is the probability. In the example, you first find -0.6 in the left column, and then .07 in the top row. Their intersection is .2514. The answer, then, is that about 25% of the purchases were under $10 (Figure 34).

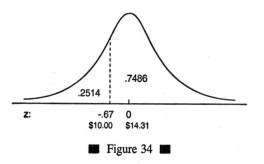

| z: | | -.67 | 0 |
| | | $10.00 | $14.31 |

.2514   .7486

■ Figure 34 ■

What if you had wanted to know the percentage of purchases above a certain amount? Because Table 2 gives the area of the curve below a given z, to obtain the area of the curve above z, simply subtract the tabled probability from 1. The area of the curve above a z of $-.67$ is $1 - .2514 = .7486$. Approximately 75% of the purchases were above $10.

Just as Table 2 can be used to obtain probabilities from z-scores, it can be used to do the reverse. For example, using the previous example, what purchase amount marks the lower 10% of the distribution?

Locate in Table 2 the probability of .1000, or as close as you can find, and read off the corresponding z-score. The figure that you seek lies between the tabled probabilities of .0985 and .1003, which correspond to z-scores of -1.29 and -1.28. To be conservative, use the more extreme value of $z = -1.29$. Now, use the z formula, this time solving for x:

$$\frac{x - 14.31}{6.4} = -1.29$$

$$x - 14.31 = (-1.29)(6.4)$$

$$x = -8.256 + 14.31 = 6.054$$

Approximately 10% of the purchases were below $6.05.

Normal Approximation to the Binomial

In Probability (page 41), you saw that some variables are continuous, that there is no limit to the number of times you could divide their intervals into still smaller ones, although you may round them off for convenience. Examples include age, height, and cholesterol level. Other variables are discrete, or made of whole units with no values between them. Some discrete variables are the number of children in a family, the sizes of televisions available for purchase, and the number of medals awarded at the Olympic Games.

The section on probability also introduced one kind of discrete variable, the binomial variable. A binomial variable can take only two values, often termed *successes* and *failures*. Examples include coin flips that come up either heads or tails, manufactured parts that either continue working past a certain point or do not, and basketball tosses that either fall through the hoop or do not.

You discovered that the outcomes of binomial trials have a frequency distribution, just as continuous variables do. The more binomial outcomes there are (for example, the more coins you flip simultaneously), the more closely the sampling distribution resembles a normal curve (Figure 35). You can take advantage of this fact and use the table of standard normal probabilities to estimate the likelihood of obtaining a given proportion of successes. You can do this by converting the test proportion to a z-score and looking up its probability in the standard normal table.

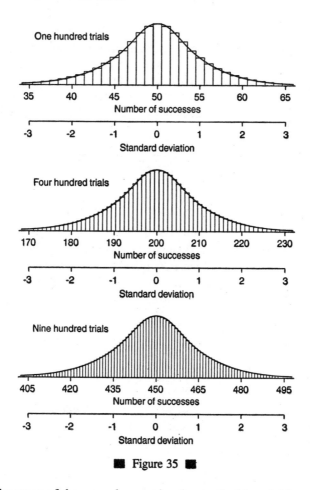

One hundred trials

Number of successes

Standard deviation

Four hundred trials

Number of successes

Standard deviation

Nine hundred trials

Number of successes

Standard deviation

■ Figure 35 ■

The mean of the normal approximation to the binomial is

$$\mu = n\pi$$

and the standard deviation is

$$\sigma = \sqrt{n\pi(1 - n\pi)}$$

where n is the number of trials and π is the total number of successes in the population divided by n. The approximation will be more accurate the larger the n, and the closer the proportion of successes in the population to .5. For example, assuming an equal chance of a new baby being a boy or a girl (i.e. $\pi = 0.5$), what is the likelihood that more than 60 out of the next 100 births at a local hospital will be boys?

$$z = \frac{x - \mu}{\sigma} = \frac{60 - ((100)(.5))}{\sqrt{(100)(.5)(1 - .5)}}$$

$$= \frac{10}{5} = 2$$

According to Table 2, a z-score of 2 corresponds to a probability of .9772. As you can see in Figure 36, there is a .9772 chance that there will be 60% or fewer boys, which means that the probability that there

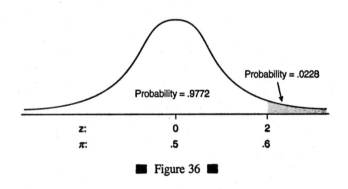

Probability = .0228

Probability = .9772

z: 0   2
π: .5   .6

■ Figure 36 ■

will be more than 60% boys is $1 - .9772 = .0228$, or just over 2%. If the assumption that the chance of a new baby being a girl is the same as it being a boy is correct, the probability of obtaining 60 or fewer girls in the next 100 births is also .9772.

Stating Hypotheses

One common use of statistics is the testing of scientific hypotheses. First, the investigator forms a **research hypothesis** that states an expectation to be tested. Then, the investigator derives a statement that is the opposite of the research hypothesis. This statement is called the **null hypothesis** (in notation: H_0). It is the null hypothesis that is actually tested, not the research hypothesis. If the null hypothesis can be rejected, that is taken as evidence in favor of the research hypothesis (also called the **alternative hypothesis,** H_a in notation). Because individual tests are rarely conclusive, it is usually not said that the research hypothesis has been "proved," only that it has been supported.

An example of a research hypothesis comparing two groups might be:

Fourth-graders in Elmwood School perform differently in math than fourth-graders in Lancaster School.

Or, in notation: $\qquad H_a: \mu_1 \neq \mu_2$

or sometimes: $\qquad H_a: \mu_1 - \mu_2 \neq 0$

The null hypothesis would be:

Fourth-graders in Elmwood School perform the same in math as fourth-graders in Lancaster School.

In notation: $\qquad H_0: \mu_1 = \mu_2$

or alternatively: $\qquad H_0: \mu_1 - \mu_2 = 0$

Some research hypotheses are more specific than that, predicting not only a difference but a difference in a particular direction:

Fourth-graders in Elmwood School are *better* in math than fourth-graders in Lancaster School.

In notation: $$H_a: \mu_1 > \mu_2$$

or alternatively: $$H_a: \mu_1 - \mu_2 > 0$$

The accompanying null hypothesis must be equally specific so that all possibilities are covered:

Fourth-graders in Elmwood School are *worse* in math, *or equal to*, fourth-graders in Lancaster School.

In notation: $$H_0: \mu_1 \le \mu_2$$

or alternatively: $$H_0: \mu_1 - \mu_2 \le 0$$

The Test Statistic

Hypothesis testing involves the use of distributions of known area, like the normal distribution, to estimate the probability of obtaining a certain value as a result of chance. The researcher is usually betting that the probability will be low because that means it's likely that the test result was not a mere coincidence, but occurred because the researcher's theory is correct. It could mean, for example, that it's probably not just bad luck but faulty packaging equipment that caused you to get a box of raisin cereal with only five raisins in it.

Only two outcomes of a hypothesis test are possible: either the null hypothesis is rejected, or it is not. You have seen that values from normally distributed populations can be converted to z-scores and their probabilities looked up in Table 2 (page 150). The z-score is one kind of **test statistic** that is used to determine the probability of obtaining a given value. In order to test hypotheses, you must decide in advance what number to use as a cutoff for whether the null hypothesis will be rejected or not. This number is sometimes called the **critical** or **tabled value** because it is looked up in a table. It represents the level of probability that you will use to test the hypothesis. If the computed test statistic has a smaller probability than that of the critical value, the null hypothesis will be rejected.

For example, suppose you want to test the theory that sunlight helps prevent depression. One hypothesis derived from this theory might be that hospital admission rates for depression in sunny regions of the country are lower than the national average. Suppose that you know the national annual admission rate for depression to be 17 per 10,000. You intend to take the mean of a sample of admission rates from hospitals in sunny parts of the country and compare it to the national average.

Your research hypothesis is:

The mean annual admission rate for depression from the hospitals in sunny areas is less than 17 per 10,000.

In notation: H_a: $\mu < 17$ per 10,000

The null hypothesis is:

The mean annual admission rate for depression from the hospitals in sunny areas is equal to or greater than 17 per 10,000.

In notation: H_0: $\mu \geq 17$ per 10,000

Your next step is to choose a probability level for the test. You know that the sample mean must be lower than 17 per 10,000 in order to reject the null hypothesis, but how much lower? You settle on a probability level of 95%. That is, if the mean admission rate for the sample of sunny hospitals is so low that the chance of obtaining that rate from a sample selected at random from the national population is less than 5%, you'll reject the null hypothesis and conclude that there is evidence to support the hypothesis that exposure to the sun reduces the incidence of depression.

Next, you look up the critical z-score—the z-score that corresponds to your chosen level of probability—in the standard normal table. It is important to remember which end of the distribution with which you are to be concerned. Table 2 (page 150) lists the probability of obtaining a given z-score or lower. That is, it gives the area of the curve below the z-score. Because a computed test statistic in the lower end of the distribution will allow you to reject your null hypothesis, you look up the z-score for the probability (or area) of .05 and find that it is −1.65. If you were hypothesizing that the mean in sunny parts of the country is greater than the national average, you would have been concerned with the upper end of the distribution instead, and would have looked up the z-score associated with the probability (area) of .95, which is $z = 1.65$.

The critical z-score allows you to define the **region of acceptance** and the **region of rejection** of the curve (Figure 37). If the computed test statistic is below the critical z-score, you can reject the null hypothesis and say that you have provided evidence in support of the alternative hypothesis. If it is above the critical value, you cannot reject the null hypothesis.

Suppose that the mean admission rate for the sample of hospitals in sunny regions is 13 per 10,000, and suppose also that the corresponding z-score for that mean is −1.20. The test statistic falls in the region of acceptance; so, you cannot reject the null hypothesis that the mean in sunny parts of the country is significantly lower than the mean in the national average. There is a greater than a 5% chance of obtaining a mean admission rate of 13 per 10,000 or lower from a sample of hospitals chosen at random from the national population, so you cannot conclude that your sample mean could not have come from that population.

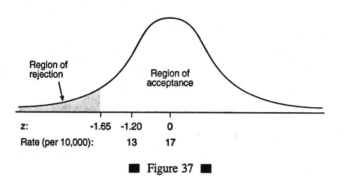

Region of
rejection

Region of
acceptance

z:		-1.65	-1.20	0
Rate (per 10,000):			13	17

■ Figure 37 ■

One- and Two-tailed Tests

In the previous example, you tested a research hypothesis that predicted not only that the sample mean would be different from the population mean but that it would be different in a specific direction—it would be lower. This test is called a **directional** or **one-tailed test** because the region of rejection is entirely within one tail of the distribution.

Some hypotheses predict only that one value will be different from another, without additionally predicting which will be higher. The test of such a hypothesis is **non-directional** or **two-tailed** because an extreme test statistic in either tail of the distribution (positive or negative) will lead to the rejection of the null hypothesis of no difference.

Suppose that you suspect that a particular class's performance on a proficiency test is not representative of those people who have taken the test. The national mean score on the test is 74.

The research hypothesis is:

The mean score of the class on the test is not 74.

Or, in notation: H_a: $\mu \neq 74$

The null hypothesis is:

The mean score of the class on the test is 74.

In notation: H_0: $\mu = 74$

As in the last example, you decide to use a 95% probability level for the test. Both tests have a region of rejection, then, of five percent, or .05. In this example, however, the rejection region must be split between both tails of the distribution—.025 in the upper tail and .025 in the lower tail (Figure 38a)—because your hypothesis specifies only a difference, not a direction. You will reject the null hypotheses of no difference if the class sample mean is either much higher or much lower than the population mean of 74. In the previous example, only a sample mean much lower than the population mean would have led to the rejection of the null hypothesis.

The decision of whether to use a one- or a two-tailed test is important because a test statistic that falls in the region of rejection in a one-tailed test may not do so in a two-tailed test, even though both tests use the same probability level. Suppose the class sample mean in your example was 77, and its corresponding z-score was computed to be 1.80. Table 2 (page 150) shows the critical z-scores for a probability of .025 in either tail to be -1.96 and 1.96. In order to reject the null hypothesis, the test statistic must be either smaller than -1.96 or greater than 1.96. It is not, so you cannot reject the null hypothesis (Figure 38(a)).

Suppose, however, you had a reason to expect that the class would perform better on the proficiency test than the population, and you did a one-tailed test, instead. For this test, the rejection region of .05 would be entirely within the upper tail. The critical z-value for a probability of .05 in the upper tail is 1.65. (Remember that the table gives areas of the curve below z; so, you look up the z-value for a probability of .95). Your computed test statistic of $z = 1.80$ exceeds the critical value and falls in the region of rejection, so you reject the null hypothesis and say that your suspicion that the class was better than the population was supported (Figure 38(b)).

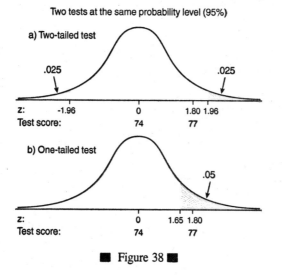

Two tests at the same probability level (95%)

a) Two-tailed test

.025 .025

z: -1.96 0 1.80 1.96
Test score: 74 77

b) One-tailed test

.05

z: 0 1.65 1.80
Test score: 74 77

■ Figure 38 ■

In practice, you should use a one-tailed test only when you have good reason to expect that the difference will be in a particular direction. A two-tailed test is more conservative than a one-tailed test—it takes a more extreme test statistic to reject the null hypothesis in a two-tailed test.

Type I and II Errors

You have been using probability to decide whether a statistical test provides evidence for or against your predictions. If the likelihood of obtaining a given test statistic from the population is very small, you reject the null hypothesis and say that you have supported your hunch that the sample you are testing is different from the population.

But you could be wrong. Even if you choose a probability level of 95%, that means there is a 5% chance, or 1 in 20, that you rejected the null hypothesis when it was, in fact, correct. You can err in the opposite way, too; you might fail to reject the null hypothesis when it is, in fact, incorrect. These two errors are called **Type I** and **Type II**, respectively. The following chart presents the four possible outcomes of any hypothesis test based on (1) whether the null hypothesis was accepted or rejected and (2) whether the null hypothesis was true in reality.

		H_0 is actually:	
		True	False
Test	reject H_0	Type I error	correct
Decision	accept H_0	correct	Type II error

Type I error is often represented by the Greek letter alpha (α) and Type II error by the Greek letter beta (β). In choosing a level of probability for a test, you are actually deciding how much you want to risk committing a Type I error—rejecting the null hypothesis when it is, in fact, true. For this reason, the area in the region of rejection is sometimes called the alpha level because it represents the likelihood of committing a Type I error.

In order to graphically depict a Type II, or β, error, it is necessary to imagine next to the distribution for the null hypothesis a second distribution for the true alternative (Figure 39). If the alternative hypothesis is actually true, but you fail to reject the null hypothesis for all values of the test statistic falling to the left of the critical value, then the area of the curve of the alternative (true) hypothesis lying to the left of the critical value represents the percentage of times that you will have made a Type II error.

Null hypothesis

True alternative

☒ Type I error
☑ Type II error
☐ Power

Critical value

■ Figure 39 ■

Type I and Type II errors are inversely related—as one increases, the other decreases. The Type I, or alpha, error rate is usually set in advance by the researcher. The Type II error rate for a given test is harder to know because it requires estimating the distribution of the alternative hypothesis, which is usually unknown.

A related concept is **power**—the probability that a test will reject the null hypothesis when it is, in fact, false. You can see from Figure 39 that power is simply 1 minus the Type II error rate (β). High power is desirable. Like β, power can be difficult to estimate accurately, but increasing the sample size always increases power.

Significance

How do you know how much confidence to put in the outcome of a hypothesis test? The statistician's criterion is the **statistical signifi-cance** of the test, or the likelihood of obtaining a given result by chance. This concept has already been spoken of, using several terms: probability, area of the curve, Type I error rate, and so forth. Another common representation of significance is the letter p (for probability) and a number between 0 and 1. There are several ways to refer to the significance level of a test, and it is important to be familiar with them. All of the following statements, for example, are equivalent:

- The finding is significant at the .05 level.
- The confidence level is 95%.
- The Type I error rate is .05.
- The alpha level is .05.
- $\alpha = .05$.
- There is a 95% certainty that the result is not due to chance.
- There is a 1 in 20 chance of obtaining this result.
- The area of the region of rejection is .05.
- The p-value is .05.
- $p = .05$.

The smaller the significance level p, the more stringent the test, and the greater the likelihood that the conclusion is correct. The significance level is usually chosen in consideration of other factors that affect and are affected by it, like sample size, estimated size of the effect being tested, and consequences of making a mistake. Common significance levels are .05 (1 chance in 20), .01 (1 chance in 100), and .001 (1 chance in 1,000).

The result of a hypothesis test, as has been seen, is that the null hypothesis is either rejected or not. The significance level for the test is set in advance by the researcher in choosing a critical test value. When the computed test statistic is large (or small) enough to reject the null hypothesis, however, it is customary to report the observed (actual) p-value for the statistic.

If, for example, you intend to perform a one-tailed (lower tail) test using the standard normal distribution at $p = .05$, the test statistic will have to be smaller than the critical z-value of -1.65 in order to reject the null hypothesis. But suppose the computed z-score is -2.50, which has an associated probability of .0062. The null hypothesis is rejected with room to spare. The observed significance level of the computed statistic is $p = .0062$; so, you could report that the result was significant at $p < .01$. This result means that even if you had chosen the more stringent significance level of .01 in advance, you would still have rejected the null hypothesis, which is stronger support for your research hypothesis than rejecting the null hypothesis at $p = .05$.

It is important to realize that statistical significance and substantive, or practical, significance are not the same thing. A small, but important, real-world difference may fail to reach significance in a statistical test. Conversely, a statistically significant finding may have no practical consequence. This finding is especially important to remember when working with large sample sizes because any difference can be statistically significant if the samples are extremely large.

Point Estimates and Confidence Intervals

You have seen that the sample mean $\bar{x}$ is an unbiased estimate of the population mean μ. Another way to say this is that $\bar{x}$ is the best **point estimate** of the true value of μ. Some error is associated with this estimate, however—the true population mean may be larger, or smaller, than the sample mean. Instead of a point estimate, you might want to identify a range of possible values μ might take, controlling the probability that μ is not lower than the lowest value in this range and not higher than the highest value. Such a range is called a **confidence interval**.

Suppose that you want to find out the average weight of all players on the football team at Landers College. You are able to select ten players at random and weigh them. The mean weight of the sample of players is 198, so that number is your point estimate. The population standard deviation is $\sigma = 11.50$. What is a 90% confidence interval for the population weight?

This question is the same as asking what weight values correspond to the upper and lower limits of an area of 90% in the center of the distribution. You can define that area by looking up in Table 2 (page 150) the z-scores that correspond to probabilities of .05 in either end of the distribution. They are -1.65 and 1.65. You can determine the weights that correspond to these z-scores using the following formula:

$$(a,b) = \bar{x} \pm z \cdot \frac{\sigma}{\sqrt{n}}$$

The weight values for the lower and upper ends of the confidence interval are 192 and 204 (Figure 40). A confidence interval is usually expressed by two values enclosed by parentheses, as in (192, 204). Another way to express the confidence interval is as the point estimate plus or minus a margin of error; in this case, it is 198 ± 6 pounds. You are 90% certain that the true population mean of football player weights is between 192 and 204 pounds.

■ Figure 40 ■

What would happen to the confidence interval if you wanted to be 95% certain of it? You would have to draw the limits (ends) of the intervals closer to the tails, in order to encompass an area of .95 between them instead of .90. That would make the low value lower and the high value higher, which would make the interval wider. The width of the confidence interval is related to the significance level, standard error, and n such that the following are true:

- The higher the percentage of accuracy (significance) desired, the wider the confidence interval.
- The larger the standard error, the wider the confidence interval.

- The larger the n, the smaller the standard error, and so the narrower the confidence interval.

All other things being equal, a smaller confidence interval is always more desirable than a larger one because a smaller interval means the population parameter can be estimated more accurately.

Estimating a Difference Score

Imagine that instead of estimating a single population mean μ, you wanted to estimate the difference between two population means μ_1 and μ_2, such as the difference between the mean weights of two football teams. The statistic $\bar{x}_1 - \bar{x}_2$ has a sampling distribution just as the individual means do, and the rules of statistical inference can be used to calculate either a point estimate or a confidence interval for the difference between the two population means.

Suppose you wanted to know which was greater, the mean weight of Landers College's football team or the mean weight of Ingram College's team. You already have a point estimate of 198 pounds for Landers' team. Suppose that you draw a random sample of players from Ingram's team, and the sample mean is 195. The point estimate for the difference between the mean weights of Landers' team (μ_1) and Ingram's team (μ_2) is $198 - 195 = 3$.

But how accurate is that estimate? You can use the sampling distribution of the difference score to construct a confidence interval for $\mu_1 - \mu_2$, just as you did for Figure 40. Suppose that when you do so, you find that the confidence interval limits are $(-3, 9)$, which means that you are 90% certain that the mean for the Landers team is between three pounds lighter and nine pounds heavier than the mean for the Ingram team.

Suppose that instead of a confidence interval, you want to test the two-tailed hypothesis that the two team weights have different means. Your null hypothesis would be:

$$H_0\colon \mu_1 = \mu_2$$

or:
$$H_0\colon \mu_1 - \mu_2 = 0$$

To reject the null hypothesis of equal means, the test statistic—in this example, z-score—for a difference in mean weights of 0 would have to fall in the rejection region at either end of the distribution. But you have already seen that it does not—only difference scores less than −3 or greater than 9 fall in the rejection region. For this reason, you would be unable to reject the null hypothesis that the two population means are equal.

This characteristic is a simple but important one of confidence intervals for difference scores. If the interval contains 0, you would be unable to reject the null hypothesis that the means are equal at the same significance level.

Univariate Tests: An Overview

To summarize, hypothesis testing of problems with one variable requires carrying out the following steps:

1. State the null hypothesis and the alternative hypothesis.
2. Decide on a significance level for the test.
3. Compute the value of a test statistic.
4. Compare the test statistic to a critical value from the appropriate probability distribution corresponding to your chosen level of significance, and observe whether the test statistic falls within the region of acceptance or the region of rejection.

Thus far, you have used the test statistic z and the table of standard normal probabilities to carry out your tests. There are other test statistics and other probability distributions. The general formula for computing a test statistic for making an inference about a single population is

$$\text{test statistic} = \frac{\text{observed sample statistic} - \text{tested value}}{\text{standard error}}$$

where *observed sample statistic* is the statistic of interest from the sample (usually the mean), *tested value* is the hypothesized population parameter (again, usually the mean), and *standard error* is the standard deviation of the sampling distribution divided by the positive square root of n.

The general formula for computing a test statistic for making an inference about a difference between two populations is

$$\text{test statistic} = \frac{\text{statistic}_1 - \text{statistic}_2 - \text{tested value}}{\text{standard error}}$$

where *statistic*$_1$ and *statistic*$_2$ are the statistics from the two samples (usually the means) to be compared, *tested value* is the hypothesized difference between the two population parameters (0 if testing for equal values), and *standard error* is the standard error of the sampling distribution, whose formula varies according to the type of problem.

The general formula for computing a confidence interval is

$$(a,b) = \bar{x} \pm \text{test statistic} \cdot \text{standard error}$$

where a and b are the lower and upper limits of the confidence interval, $\bar{x}$ is the point estimate (usually the sample mean), *test statistic* is the critical value from the table of the appropriate probability distribution (upper or positive value if z) corresponding to half of the desired alpha level, and *standard error* is the standard error of the sampling distribution.

Why must the alpha level be halved before looking up the critical value when computing a confidence interval? Because the rejection region is split between both tails of the distribution, as in a two-tailed test. For a confidence interval at $\alpha = .05$, then, you would look up the critical value corresponding to an upper-tailed probability of .025.

One-sample z-test

Requirements: normally distributed population, σ known

Test for population mean.

Hypothesis test

> **Formula:**

$$z = \frac{\bar{x} - \Delta}{\dfrac{\sigma}{\sqrt{n}}}$$

where $\bar{x}$ is the sample mean, Δ is a specified value to be tested, σ is the population standard deviation, and n is the size of the sample. Look up the significance level of the z-value in the standard normal table (Table 2, page 150).

Example (one-tailed test): A herd of 1,500 steers was fed a special high-protein grain for a month. A random sample of 29 were weighed and had gained an average of 6.7 pounds. If the standard deviation of weight gain for the entire herd is 7.1, what is the likelihood that the average weight gain per steer for the month was at least 5 pounds?

null hypothesis: H_0: $\mu < 5$

alternative hypothesis: H_a: $\mu \geq 5$

$$z = \frac{6.7 - 5}{\frac{7.1}{\sqrt{29}}} = \frac{1.7}{1.318} = 1.289$$

tabled value for $z \leq 1.28$ is .8997

$1 - .8997 = .1003$

So, the probability that the herd gained at least 5 pounds per steer is $p < .1003$. Should the null hypothesis of a weight gain of less than 5 pounds for the population be rejected? That depends on how conservative you want to be. If you had decided beforehand on a significance level of $p < .05$, the null hypothesis could not be rejected.

Example (two-tailed test): In national use, a vocabulary test is known to have a mean score of 68 and a standard deviation of 13. A class of 19 students takes the test and has a mean score of 65. Is the class typical of others who have taken the test? Assume a significance level of $p < .05$.

There are two possible ways that the class may differ from the population. Its scores may be lower than, or higher than, the population of all students taking the test; therefore, this problem requires a two-tailed test. First, state the null and alternative hypotheses:

null hypothesis: H_0: $\mu = 68$

alternative hypothesis: H_a: $\mu \neq 68$

Because you have specified a significance level, you can look up the critical z-value in the table before computing the statistic. This is a two-tailed test; so, the .05 must be split such that .025 is in the upper tail and another .025 in the lower. The z-value that corresponds to –.025 is –1.96, which is the lower critical z-value. The upper value corresponds to $1 - .025$, or .975, which gives a z-value of 1.96. The null hypothesis of no difference will be rejected if the computed z statistic falls outside of the range of –1.96 to 1.96.

Next, compute the z statistic:

$$z = \frac{65 - 68}{\dfrac{13}{\sqrt{19}}} = \frac{-3}{2.982} = -1.006$$

Because –1.006 is between –1.96 and 1.96, the null hypothesis of population mean is 68 cannot be rejected. That is, the class can be considered as typical of others who have taken the test.

Confidence interval for population mean using z.

Formula:

$$(a,b) = \bar{x} \pm z_{\alpha/2} \cdot \frac{\sigma}{\sqrt{n}}$$

where a and b are the limits of the confidence interval, $\bar{x}$ is the sample mean, $z_{\alpha/2}$ is the upper (or positive) z-value from the standard normal table corresponding to half of the desired alpha level (because all confidence intervals are two-tailed), σ is the population standard deviation, and n is the size of the sample.

Example: A sample of 12 machine pins has a mean diameter of 1.15 inches, and the population standard deviation is known to be .04. What is a 99% confidence interval of diameter width for the population?

First, determine the z-value. A 99% confidence level is equivalent to $p < .01$. Half of .01 is .005. The z-value corresponding to an area of .005 is 2.58. The interval may now be calculated:

$$(a,b) = 1.15 \pm 2.58 \cdot \frac{.04}{\sqrt{12}}$$

$$= 1.15 \pm .03$$

$$= (1.12, \ 1.18)$$

It is 99% certain that the population mean of pin diameters lies between 1.12 and 1.18 inches. Note that this is not the same as saying that 99% of the machine pins have diameters between 1.12 and 1.18 inches, which would be an incorrect conclusion from this test.

Choosing a sample size. Because surveys cost money to administer, researchers often want to calculate how many subjects will be needed to determine a population mean using a fixed confidence interval and significance level. The formula is

$$n = \left(\frac{2 \, z_{\alpha/2} \sigma}{w} \right)^2$$

where n is the number of subjects needed, $z_{\alpha/2}$ is the critical z-value corresponding to the desired significance level, σ is the population standard deviation, and w is the desired confidence interval width.

Example: How many subjects will be needed to find the average age of students at Fisher College plus or minus a year, with a 95% significance level and a population standard deviation of 3.5?

$$n = \left(\frac{(2)(1.96)(3.5)}{2}\right)^2 = \left(\frac{13.72}{2}\right)^2 = 47.06$$

Rounding up, a sample of 48 students would be sufficient to determine students' mean age plus or minus one year. Note that the confidence interval width is always double the "plus or minus" figure.

One-sample t-test

Requirements: normally distributed population, σ is unknown

Test for population mean.

Hypothesis test

Formula:
$$t = \frac{\bar{x} - \Delta}{\dfrac{s}{\sqrt{n}}}$$

where $\bar{x}$ is the sample mean, Δ is a specified value to be tested, s is the sample standard deviation (also called standard error of the mean), and n is the size of the sample. When the standard deviation of the sample is substituted for the standard deviation of the population, the statistic does not have a normal distribution—it has what is called the t-distribution (Table 3, page 154). Because there is a different t-distribution for each sample size, it is not practical to list a separate area-of-the-curve table for each one. Instead, critical t-values for common alpha levels (.05, .01, .001, and so forth) are usually given in a single table for a range of sample sizes. For very large samples, the t-distribution approximates the standard normal (z) distribution.

Values in the t-table are not actually listed by sample size, but by **degrees of freedom** (df). The number of degrees of freedom for a problem involving the t-distribution for sample size n is simply $n - 1$ for a one-sample mean problem.

Example (one-tailed test): A professor wants to know if her introductory statistics class has a good grasp of basic math. Six students are chosen at random from the class and given a math proficiency test. The professor wants the class to be able to score at least 70 on the test. The six students get scores of 62, 92, 75, 68, 83, and 95. Can the professor be at least 90% certain that the mean score for the class on the test would be at least 70?

null hypothesis: H_0: $\mu < 70$

alternative hypothesis: H_a: $\mu \geq 70$

First, compute the sample mean and standard deviation (see pages 29 and 38).

$$
\begin{array}{l}
62 \\
92 \\
75 \\
68 \\
83 \\
95 \\
\hline
475
\end{array}
\qquad
\bar{x} = \frac{475}{6} = 79.17
$$

$$s = 13.17$$

Next, compute the t-value:

$$t = \frac{79.17 - 70}{\dfrac{13.17}{\sqrt{6}}} = \frac{9.17}{5.38} = 1.71$$

To test the hypothesis, the computed t-value of 1.71 will be compared to the critical value in the t-table. But which do you expect to be larger and which smaller? One way to reason about this is to look at the formula and see what effect different means would have on the computation. If the sample mean had been 85 instead of 79.17, the resulting t-value would have been larger. Because the sample mean is in the numerator, the larger it is, the larger the resulting figure will be. At the same time, you know that a higher sample mean will make it more likely that the professor will conclude that the math proficiency of the class is satisfactory and that the null hypothesis of less-than-satisfactory class math knowledge can be rejected. Therefore, it must be true that the larger the computed t-value, the greater the chance that the null hypothesis can be rejected. It follows, then, that if the computed t-value is larger than the critical t-value from the table, the null hypothesis can be rejected.

A 90% confidence level is equivalent to an alpha level of .10. Because extreme values in one rather than two directions will lead to rejection of the null hypothesis, this is a one-tailed test, and you do not divide the alpha level by 2. The number of degrees of freedom for the problem is $6 - 1 = 5$. The value in the t-table for $t_{.10, 5}$ is 1.476. Because the computed t-value of 1.71 is larger than the critical value in the table, the null hypothesis can be rejected, and the professor can be 90% certain that the class mean on the math test would be at least 70.

Note that the formula for the one-sample t-test for a population mean is the same as the z-test, except that the t-test substitutes the sample standard deviation s for the population standard deviation σ and takes critical values from the t-distribution instead of the z-distribution. The t-distribution is particularly useful for tests with small samples.

Example (two-tailed test): A Little League baseball coach wants to know if his team is representative of other teams in scoring runs. Nationally, the average number of runs scored by a Little League team in a game is 5.7. He chooses five games at random in which his team scored 5, 9, 4, 11, and 8 runs. Is it likely that his team's scores could have come from the national distribution? Assume an alpha level of .05.

Because the team's scoring rate could be either higher than or lower than the national average, the problem calls for a two-tailed test. First, state the null and alternative hypotheses:

null hypothesis: H_0: $\mu = 5.7$

alternative hypothesis: H_a: $\mu \neq 5.7$

Next, compute the sample mean and standard deviation:

$$
\begin{array}{l}
5 \\
9 \\
4 \\
11 \\
\underline{8} \\
37
\end{array}
\qquad
\begin{array}{l}
\bar{x} = \dfrac{37}{5} = 7.4 \\[2mm]
s = 2.88
\end{array}
$$

Next, the t-value:

$$t = \frac{7.4 - 5.7}{\dfrac{2.88}{\sqrt{5}}} = \frac{1.7}{1.29} = 1.32$$

Now, look up the critical value from the t-table. You need to know two things in order to do this: the degrees of freedom and the desired alpha level. The degrees of freedom is $5 - 1 = 4$. The overall alpha level is .05, but because this is a two-tailed test, the alpha level must be divided by two, which yields .025. The tabled value for $t_{.025,4}$ is 2.776. The computed t of 1.32 is smaller than the t from the table; so, you cannot reject the null hypothesis that the mean of this team is equal to the population mean. The coach can conclude that his team fits in with the national distribution on runs scored.

Confidence interval for population mean using t.

Formula: $$(a,b) = \bar{x} \pm t_{\alpha/2,df} \cdot \frac{s}{\sqrt{n}}$$

where a and b are the limits of the confidence interval, $\bar{x}$ is the sample mean, $t_{\alpha/2,df}$ is the value from the t-table corresponding to half of the desired alpha level at $n-1$ degrees of freedom, s is the sample standard deviation, and n is the size of the sample.

Example: Using the previous example, what is a 95% confidence interval for runs scored per team per game?

First, determine the t-value. A 95% confidence level is equivalent to an alpha level of .05. Half of .05 is .025. The t-value corresponding to an area of .025 at either end of the t-distribution for 4 degrees of freedom ($t_{.025,4}$) is 2.776. The interval may now be calculated:

$$(a,b) = 5.7 \pm 2.78 \frac{2.88}{\sqrt{5}}$$

$$= 5.7 \pm 3.58$$

$$= (2.12, 9.28)$$

The interval is fairly wide, mostly because n is small.

Two-sample z-test for Comparing Two Means

Requirements: two normally distributed but independent populations, σ is known

Hypothesis test

Formula:
$$z = \frac{\bar{x}_1 - \bar{x}_2 - \Delta}{\sqrt{\dfrac{\sigma_1^2}{n_1} + \dfrac{\sigma_2^2}{n_2}}}$$

where $\bar{x}_1$ and $\bar{x}_2$ are the means of the two samples, Δ is the hypothesized difference between the population means (0 if testing for equal means), σ_1 and σ_2 are the standard deviations of the two populations, and n_1 and n_2 are the sizes of the two samples.

Example (two-tailed test): The amount of a certain trace element in blood is known to vary with a standard deviation of 14.1 ppm (parts per million) for male blood donors and 9.5 ppm for female donors. Random samples of 75 male and 50 female donors yield concentration means of 28 and 33 ppm, respectively. What is the likelihood that the population means of concentrations of the element are the same for men and women?

null hypothesis: $H_0: \mu_1 = \mu_2$

or: $H_0: \mu_1 - \mu_2 = 0$

alternative hypothesis: $H_a: \mu_1 \neq \mu_2$

or: $H_a: \mu_1 - \mu_2 \neq 0$

$$z = \cfrac{28 - 33 - 0}{\sqrt{\cfrac{14.1^2}{75} + \cfrac{9.5^2}{50}}} = \frac{-5}{\sqrt{2.65 + 1.81}} = -2.37$$

The computed z-value is negative because the (larger) mean for females was subtracted from the (smaller) mean for males. But because the hypothesized difference between the populations is 0, the order of the samples in this computation is arbitrary—$\bar{x}_1$ could just as well have been the female sample mean and $\bar{x}_2$ the male sample mean, in which case z would be 2.37 instead of −2.37. An extreme z-score in either tail of the distribution (plus or minus) will lead to rejection of the null hypothesis of no difference.

The area of the standard normal curve corresponding to a z-score of −2.37 is .0089. Because this test was two-tailed, that figure is doubled to yield a probability of .0178 that the population means are the same. If the test had been conducted at a pre-specified significance level of $p < .05$, the null hypothesis of equal means could be rejected. If the specified significance level had been the more conservative (more stringent) $p < .01$, however, the null hypothesis could not be rejected.

In practice, the two-sample z-test is not often used because the two population standard deviations σ_1 and σ_2 are usually unknown. Instead, sample standard deviations and the t-distribution are used.

Two-sample *t*-test for Comparing Two Means

Requirements: two normally distributed but independent populations, σ is unknown

Hypothesis test

Formula:
$$t = \frac{\bar{x}_1 - \bar{x}_2 - \Delta}{\sqrt{\dfrac{s_1^2}{n_1} + \dfrac{s_2^2}{n_2}}}$$

where $\bar{x}_1$ and $\bar{x}_2$ are the means of the two samples, Δ is the hypothesized difference between the population means (0 if testing for equal means), s_1 and s_2 are the standard deviations of the two samples, and n_1 and n_2 are the sizes of the two samples. The number of degrees of freedom for the problem is the smaller of $n_1 - 1$ and $n_2 - 1$.

Example (one-tailed test): An experiment is conducted to determine whether intensive tutoring (covering a great deal of material in a fixed amount of time) is more effective than paced tutoring (covering less material in the same amount of time). Two randomly chosen groups are tutored separately and then administered proficiency tests. Use a significance level of $p < .05$.

null hypothesis: $H_0: \mu_1 \leq \mu_2$

or: $H_0: \mu_1 - \mu_2 \leq 0$

alternative hypothesis: $H_a: \mu_1 > \mu_2$

or: $$H_a: \mu_1 - \mu_2 > 0$$

Group	Method	n	$\bar{x}$	s
1	intensive	12	46.31	6.44
2	paced	10	42.79	7.52

$$t = \frac{46.31 - 42.79 - 0}{\sqrt{\dfrac{6.44^2}{12} + \dfrac{7.52^2}{10}}} = \frac{3.52}{\sqrt{3.46 + 5.66}} = 1.166$$

The degrees of freedom parameter is the smaller of $(12 - 1)$ and $(10 - 1)$, or 9. Because this is a one-tailed test, the alpha level (.05) is not divided by two. The next step is to look up $t_{.05,9}$ in the t-table, which gives a critical value of 1.833. The computed t of 1.166 does not exceed the tabled value, so the null hypothesis cannot be rejected. This test has not provided statistically significant evidence that intensive tutoring is superior to paced tutoring.

Confidence interval for comparing two means.

Formula:

$$(a,b) = \bar{x}_1 - \bar{x}_2 \pm t_{\alpha/2, df} \cdot \sqrt{\frac{s_1^2}{n_1} + \frac{s_2^2}{n_2}}$$

where a and b are the limits of the confidence interval, $\bar{x}_1$ and $\bar{x}_2$ are the means of the two samples, $t_{\alpha/2, df}$ is the value from the t-table corresponding to half of the desired alpha level, s_1 and s_2

are the standard deviations of the two samples, and n_1 and n_2 are the sizes of the two samples. The degrees of freedom parameter for looking up the t-value is the smaller of $n_1 - 1$ and $n_2 - 1$.

Example: Estimate a 90% confidence interval for the difference between the number of raisins per box in two brands of breakfast cereal.

Brand	n	$\bar{x}$	s
A	6	102.1	12.3
B	9	93.6	7.52

The difference between $\bar{x}_1$ and $\bar{x}_2$ is $102.1 - 93.6 = 8.5$. The degrees of freedom is the smaller of $(6 - 1)$ and $(9 - 1)$, or 5. A 90% confidence interval is equivalent to an alpha level of .10, which is then halved to give .05. According to the table, the critical value for $t_{.05,5}$ is 2.015. The interval may now be computed.

$$(a,b) = 8.5 \pm 2.015 \cdot \sqrt{\frac{(12.3)^2}{6} + \frac{(7.52)^2}{9}}$$

$$= 8.5 \pm 2.015 \cdot \sqrt{25.22 + 6.28}$$

$$= 8.5 \pm 11.31$$

$$= (-2.81, 19.81)$$

You can be 90% certain that Brand A cereal has between 2.81 fewer and 19.81 more raisins per box than Brand B. The fact that the interval contains 0 means that if you had performed a test of the hypothesis that the two population means are different (using the same significance level), you would not have been able to reject the null hypothesis of no difference.

Pooled variance method. If the two population distributions can be assumed to have the same variance—and therefore the same standard deviation—s_1 and s_2 can be pooled together, each weighted by the number of cases in each sample. Although using pooled variance in a t-test is generally more likely to yield significant results than using separate variances, it is often hard to know whether the variances of the two populations are equal. For this reason, the pooled variance method should be used with caution. The formula for the pooled estimator of σ^2 is

$$S_p^2 = \frac{(n_1 - 1) s_1^2 + (n_2 - 1) s_2^2}{n_1 + n_2 - 2}$$

where s_1 and s_2 are the standard deviations of the two samples and n_1 and n_2 are the sizes of the two samples.

The formula for comparing the means of two populations using pooled variance is

$$t = \frac{\bar{x}_1 - \bar{x}_2 - \Delta}{\sqrt{S_p^2 \left(\dfrac{1}{n_1} + \dfrac{1}{n_2} \right)}}$$

where $\bar{x}_1$ and $\bar{x}_2$ are the means of the two samples, Δ is the hypothesized difference between the population means (0 if testing for equal means), s_p^2 is the pooled variance, and n_1 and n_2 are the sizes of the two samples. The number of degrees of freedom for the problem is

$$df = n_1 + n_2 - 2$$

Example (two-tailed test): Does right- or left-handedness affect how fast people type? Random samples of students from a typing class are given a typing speed test (words per minute), and the results are compared. Significance level for the test: .10. Because you are looking for a difference between the groups in either direction (right-handed faster than left, or vice versa), this is a two-tailed test.

null hypothesis: $H_0: \mu_1 = \mu_2$

or: $H_0: \mu_1 - \mu_2 = 0$

alternative hypothesis: $H_a: \mu_1 \neq \mu_2$

or: $H_a: \mu_1 - \mu_2 \neq 0$

Group	-handed	n	$\bar{x}$	s
1	right	16	55.8	5.7
2	left	9	59.3	4.3

First, calculate the pooled variance:

$$S_p^2 = \frac{(16 - 1)\,5.7^2 + (9 - 1)\,4.3^2}{16 + 9 - 2}$$

$$= \frac{487.35 + 147.92}{23}$$

$$= 27.62$$

Next, calculate the t-value:

$$t = \frac{55.8 - 59.3 - 0}{\sqrt{27.62\left(\frac{1}{16} + \frac{1}{9}\right)}} = \frac{-3.5}{\sqrt{4.80}} = -1.598$$

The degrees-of-freedom parameter is $16 + 9 - 2$, or 23. This test is a two-tailed one; so, you divide the alpha level (.10) by two. Next, you look up $t_{.05,23}$ in the t-table (page 154), which gives a critical value of 1.714. This value is larger than the absolute value of the computed t of -1.598, so the null hypothesis of equal population means cannot be rejected. There is no evidence that right- or left-handedness has any effect on typing speed.

Paired Difference t-test

Requirements: a set of paired observations from a normal population

This t-test compares one set of measurements with a second set from the same sample. It is often used to compare "before" and "after" scores in experiments to determine whether significant change has occurred.

Hypothesis test

Formula: $$t = \frac{\bar{x} - \Delta}{\dfrac{s}{\sqrt{n}}}$$

where $\bar{x}$ is the mean of the change scores, Δ is the hypothesized difference (0 if testing for equal means), s is the sample standard deviation, and n is the sample size. The number of degrees of freedom for the problem is $n - 1$.

Example (one-tailed test): A farmer decides to try out a new fertilizer on a test plot containing 10 stalks of corn. Before applying the fertilizer, he measures the height of each stalk. Two weeks later, he measures the stalks again, being careful to match each stalk's new height to its previous one. The stalks would have grown an average of six inches during that time even without the fertilizer. Did the fertilizer help? Use a significance level of .05.

null hypothesis: H_0: $\Delta \le 6$

alternative hypothesis: H_a: $\Delta > 6$

Stalk	1	2	3	4	5	6	7	8	9	10
Before height	35.5	31.7	31.2	36.3	22.8	28.0	24.6	26.1	34.5	27.7
After height	45.3	36.0	38.6	44.7	31.4	33.5	28.8	35.8	42.9	35.0

Subtract each stalk's "before" height from its "after" height to get the change score for each stalk; then, compute the mean and standard deviation of the change scores, and insert these into the formula.

$$45.3 - 35.5 = 9.8$$
$$36.0 - 31.7 = 4.3$$
$$38.6 - 31.2 = 7.4 \qquad s = 2.05$$
$$44.7 - 36.3 = 8.4$$
$$31.4 - 22.8 = 8.6 \qquad t = \frac{7.36 - 6}{\dfrac{2.05}{\sqrt{10}}} = \frac{1.36}{.65} = 2.098$$
$$33.5 - 28.0 = 5.5$$
$$28.8 - 24.6 = 4.2$$
$$35.8 - 26.1 = 9.7$$
$$42.9 - 34.5 = 8.4$$
$$35.0 - 27.7 = 7.3$$

$$73.6 / 10 = 7.36$$

The problem has $n - 1$, or $10 - 1 = 9$ degrees of freedom. The test is one-tailed because you are asking only whether the fertilizer increases growth, not reduces it. The critical value from the t-table for $t_{.05,9}$ is 1.833. Because the computed t-value of 2.098 is larger than 1.833, the null hypothesis can be rejected. The test has provided evidence that the fertilizer caused the corn to grow more than if it had not been fertilized. The amount of actual increase was not large (1.36 inches over normal growth), but it was statistically significant.

Test for a Single Population Proportion

Requirements: binomial population, sample $n\pi_0 \geq 10$, and sample $n(1 - \pi_0) \geq 10$, where π_0 is the hypothesized proportion of successes in the population.

Hypothesis test

Formula:
$$z = \frac{\hat{\pi} - \pi_0}{\sqrt{\dfrac{\pi_0 (1 - \pi_0)}{n}}}$$

where $\hat{\pi}$ is the sample proportion, π_0 is the hypothesized proportion, and n is the sample size. Because the distribution of sample proportions is approximately normal for large samples, the z statistic is used. The test is most accurate when π (the population proportion) is close to .5 and least accurate when π is close to 0 or 1.

Example (one-tailed test): The sponsors of a city marathon have been trying to encourage more women to participate in the event. A sample of 70 runners is taken, of which 32 are women. The sponsors would like to be 90% certain that at least 40% of the participants are women. Were their recruitment efforts successful?

null hypothesis: $H_0: \pi < .4$

alternative hypothesis: $H_a: \pi \geq .4$

The proportion of women runners in the sample is 32 out of 70, or 45.7%. The z-value may now be calculated:

$$z = \frac{.457 - .40}{\sqrt{\dfrac{.40(1 - .40)}{70}}} = \frac{.057}{\sqrt{.003}} = .973$$

From the z-table, you find that the probability of a z-value less than .97 is .834, not quite the .90 required to reject the null hypothesis so that it cannot be concluded at that level of significance that the population of runners is at least 40% women.

Confidence interval for a single population proportion.

Formula: $\qquad \hat{\pi} \pm z_{\alpha/2} \cdot \sqrt{\dfrac{\hat{\pi}(1 - \hat{\pi})}{n}}$

where $\hat{\pi}$ is the sample proportion, $z_{\alpha/2}$ is the upper z-value corresponding to half of the desired alpha level, and n is the sample size.

Example: A sample of 100 voters selected at random in a congressional district prefer Candidate Smith to Candidate Jones by a ratio of 3 to 2. What is a 95% confidence interval of the percentage of voters in the district who prefer Smith?

A ratio of 3 to 2 is equivalent to a proportion of $3/5 = .60$. A 95% confidence interval is equivalent to an alpha level of .05, half of which is .025. The critical z-value corresponding to an upper probability of $1 - .025$ is 1.96. The interval may now be computed:

$$(a,b) = .60 \pm 1.96 \sqrt{\frac{.60\,(1 - .60)}{100}}$$

$$= .60 \pm 1.96 \sqrt{.002}$$

$$= .60 \pm .096$$

$$= (.504, .696)$$

It is 95% certain that between 50.4% and 69.6% of the voters in the district prefer Candidate Smith. Note that the problem could have been figured for Candidate Jones by substituting the proportion .40 for Smith's proportion of .60.

Choosing a sample size. In the previous problem, you estimated that the percentage of voters in the district who prefer Candidate Smith is 60%, plus or minus about 10%. Another way to say this is that the estimate has a "margin of error" of ±10%, or a confidence interval width of 20%. That's a pretty wide range. You may wish to make the margin smaller.

Because the width of the confidence interval decreases at a known rate as the sample size increases, it is possible to determine the sample size needed to estimate a proportion with a fixed confidence interval. The formula is

$$n = \left(\frac{2\,z_{\alpha/2}}{w}\right)^2 \cdot p* (1 - p*)$$

where n is the number of subjects needed, $z_{\alpha/2}$ is the z-value corresponding to half of the desired significance level, w is the desired confidence interval width, and $p*$ is an estimate of the true population proportion. A $p*$ of .50 will result in a higher n than any other proportion estimate but is often used when the true proportion is not known.

Example: How large a sample is needed to estimate the preference of district voters for Candidate Smith with a margin of error of ±4%, at a 95% significance level?

You will conservatively estimate the (unknown) true population proportion of preference for Smith at .50. If it's really larger (or smaller) than that, you will overestimate the size of the sample needed, but $p* = .50$ is playing it safe.

$$n = \left(\frac{2 \cdot 1.96}{.08}\right)^2 \cdot .50\,(1 - .50)$$

$$= 49^2 \cdot .25$$

$$= 600.25$$

A sample of about 600 voters would be needed to estimate the percentage of voters in the district who prefer Smith and to be 95% certain that the estimate is within (plus or minus) 4% of the true population percentage.

Test for Comparing Two Proportions

Requirements: two binomial populations, $n\pi_0 \geq 5$ and $n(1 - \pi_0) \geq 5$ (for each sample), where π_0 is the hypothesized proportion of successes in the population.

Difference test.

Hypothesis test

Formula:
$$z = \frac{\hat{\pi}_1 - \hat{\pi}_2 - \Delta}{\sqrt{\hat{\pi}(1 - \hat{\pi})\left(\dfrac{1}{n_1} + \dfrac{1}{n_2}\right)}}$$

where
$$\hat{\pi} = \frac{x_1 + x_2}{n_1 + n_2}$$

and where $\hat{\pi}_1$ and $\hat{\pi}_2$ are the sample proportions, Δ is their hypothesized difference (0 if testing for equal proportions), n_1 and n_2 are the sample sizes, and x_1 and x_2 are the number of "successes" in each sample. As in the test for a single proportion, the z-distribution is used to test the hypothesis.

Example (one-tailed test): A swimming school wants to determine whether an instructor they hired recently is working out. Sixteen out of 25 of Instructor A's students passed the lifeguard certification test on the first try. In comparison, 57 out of 72 of more experienced Instructor B's students passed the test on the first try. Is Instructor A's success rate worse than Instructor B's? Use $\alpha = .10$.

null hypothesis: H_0: $\pi_1 \geq \pi_2$

alternative hypothesis: H_a: $\pi_1 < \pi_2$

First, you need to compute the values for some of the terms in the formula. The sample proportion $\hat{\pi}_1$ is $16/25 = .640$. The sample proportion $\hat{\pi}_2$ is $57/72 = .792$. Next, compute $\hat{\pi}$:

$$\hat{\pi} = \frac{16 + 57}{25 + 72} = \frac{73}{97} = .753$$

Finally, the main formula:

$$z = \frac{.640 - .792 - 0}{\sqrt{.753\,(1 - .753)\left(\dfrac{1}{25} + \dfrac{1}{72}\right)}}$$

$$= \frac{-.152}{\sqrt{(.186)(.054)}} = \frac{-.152}{.100} = -1.517$$

The standard normal (z) table shows that the lower critical z-value for $\alpha = .10$ is approximately -1.28. The computed z must be lower than -1.28 in order to reject the null hypothesis of equal proportions. Because the computed z is -1.517, the null hypothesis can be rejected. It can be concluded (at this level of significance) that Instructor A's success rate is worse than Instructor B's.

Confidence interval for comparing two proportions.

Formula: $(a,b) = \hat{\pi}_1 - \hat{\pi}_2 \pm z_{\alpha/2} \cdot s(D)$

$$\text{where} \quad s(D) = \sqrt{\frac{\hat{\pi}_1(1 - \hat{\pi}_1)}{n_1} + \frac{\hat{\pi}_2(1 - \hat{\pi}_2)}{n_2}}$$

and where a and b are the limits of the confidence interval of $\pi_1 - \pi_2$, $\hat{\pi}_1$ and $\hat{\pi}_2$ are the sample proportions, $z_{\alpha/2}$ is the upper z-value corresponding to half of the desired alpha level, and n_1 and n_2 are the sizes of the two samples.

Example: A public health researcher wants to know how two high schools, one in the inner city and one in the suburbs, differ in the percentage of students who smoke. A random survey of students gives the following results:

Population	n	Smokers
1 (inner-city)	125	47
2 (suburban)	153	52

What is a 90% confidence interval for the difference between the smoking rates in the two schools?

The proportion of smokers in the inner-city school is $\hat{\pi}_1$ = 47/125 = .376. The proportion of smokers in the suburban school is $\hat{\pi}_2$ = 52/153 = .340. Next, solve for $s(D)$:

$$s(D) = \sqrt{\frac{.376(1 - .376)}{125} + \frac{.340(1 - .340)}{153}}$$

$$= \sqrt{\frac{.235}{125} + \frac{.224}{153}} = \sqrt{.003} = .058$$

A 90% confidence interval is equivalent to $\alpha = .10$, which is halved to give .05. The upper tabled value for $z_{.05}$ is 1.65. The interval may now be computed:

$$(a,b) = (.376 - .340) \pm (1.65)(.058)$$
$$= .036 \pm .096$$
$$= (-.060, .132)$$

The researcher can be 90% certain that the true population proportion of smokers in the inner-city high school is between 6% lower and 13.2% higher than the proportion of smokers in the suburban high school.

Thus far, we have been working with problems having a single variable. We now move on to two-variable, or **bivariate**, problems.

Correlation

Consider the following data table, which contains measurements on two variables for ten people: the number of months the person has owned an exercise machine and the number of hours the person spent exercising in the past week.

Person	1	2	3	4	5	6	7	8	9	10
Months owned	5	10	4	8	2	7	9	6	1	12
Hours exercised	5	2	8	3	8	5	5	7	10	3

If you display these data pairs as points in a scatterplot (Figure 42), you can see a definite trend. The points appear to form a line that slants from the upper left to the lower right. As you move along that line from left to right, the values on the vertical axis (hours of exercise) get smaller, while the values on the horizontal axis (months owned) get larger. Another way to express this is to say that the two variables are inversely related: the more months the machine was owned, the less the person tends to exercise.

■ Figure 42 ■

These two variables are **correlated**. More than that, they are correlated in a particular direction—negatively. For an example of a positive correlation, suppose that instead of displaying "hours of exercise" on the vertical axis, you put the person's score on a test that measures cardiovascular fitness (Figure 43). The pattern of these data points suggests a line that slants from lower left to upper right, which is the opposite of the direction of slant in the first example. Figure 43 shows that the longer the person has owned the exercise machine, the better his or her cardiovascular fitness tends to be. This might be true in spite of the fact that time spent exercising decreases the longer the machine has been owned because purchasers of exercise machines might be starting from a point of low fitness, which may improve only gradually.

■ Figure 43 ■

If two variables are positively correlated, as the value of one increases, so does the value of the other. If they are negatively (or inversely) correlated, as the value of one increases, the value of the other decreases.

A third possibility remains: that as the value of one variable increases, the value of the other neither increases nor decreases. Figure 44 is a scatterplot of months exercise machine owned (horizontal axis) by person's height (vertical axis). No line trend can be seen in the plot. New owners of exercise machines may be short or tall, and the same is true of people who have had their machines longer. These two variables appear to be uncorrelated.

Months Machine Owned

■ Figure 44 ■

You can go even farther in expressing the relationship between variables. Compare the two scatterplots in Figure 44. Both plots show a positive correlation because as the values on one axis increase, so do the values on the other. But the data points in Figure 45(b) are more closely packed than the data points in Figure 45(a), which are more spread out. If a line were drawn through the middle of the trend, the points in Figure 45(b) would be closer to the line than the points in Figure 45(a). In addition to direction (positive or negative), correlations can also have strength, which is a reflection of the closeness of the data points to a perfect line. Figure 45(b) shows a stronger correlation than Figure 45(a).

(a)   (b)

■ Figure 45 ■

Pearson's product moment coefficient (r), commonly referred to as the **correlation coefficient,** is a quantitative measure of correlation between two interval-level variables. The coefficient r can take values from -1.0 to 1.0. The sign of r indicates whether the correlation is positive or negative. The magnitude (absolute value) of r indicates the strength of the correlation, or how close the array of data points is to a straight line.

The computing formula for r is

$$r_{x,y} = \frac{\Sigma xy - \dfrac{1}{n}(\Sigma x)(\Sigma y)}{(n-1)s_x s_y}$$

where Σxy is the sum of the xy cross-products (each x multiplied by its paired y), n is the size of the sample (the number of data pairs), Σx and Σy are the sums of the x and y values, and s_x and s_y are the sample standard deviations of x and y.

Use the first table in this section (page 111) to compute r for the relationship between months of exercise-machine ownership and hours of exercise per week. The first step is to compute the components required in the main formula. Let x be months of ownership and y be hours of exercise, although you could also do the reverse.

x	y	xy	x^2	y^2
5	5	25	25	25
10	2	20	100	4
4	8	32	16	64
8	3	24	64	9
2	8	16	4	64
7	5	35	49	25
9	5	45	81	25
6	7	42	36	49
1	10	10	1	100
12	3	36	144	9
$\Sigma x = 64$	$\Sigma y = 56$	$\Sigma xy = 285$	$\Sigma x^2 = 520$	$\Sigma y^2 = 374$

$$(\Sigma x)^2 = 64^2 = 4096$$

$$(\Sigma y)^2 = 56^2 = 3136$$

Now, compute the sample standard deviations for x and y using the formula from Numerical Measures (page 38):

$$s_x = \sqrt{\frac{\Sigma x^2 - \dfrac{(\Sigma x)^2}{n}}{n-1}}$$

$$= \sqrt{\frac{520 - \dfrac{4096}{10}}{10-1}}$$

$$= \sqrt{\frac{110.4}{9}} = \sqrt{12.267} = 3.502$$

$$s_y = \sqrt{\frac{374 - \dfrac{3136}{10}}{10 - 1}}$$

$$= \sqrt{\frac{60.4}{9}} = \sqrt{6.711} = 2.591$$

Finally, r may be computed:

$$r_{x,y} = \frac{\Sigma_{xy} - \dfrac{1}{n}(\Sigma x)(\Sigma y)}{(n - 1)\,s_x\,s_y}$$

$$= \frac{285 - \dfrac{1}{10}(64)(56)}{(10 - 1)(3.502)(2.591)}$$

$$= \frac{-73.4}{81.663} = -0.899$$

A correlation of $r = -.899$ is almost as strong as the maximum negative correlation of -1.0, reflecting the fact that your data points fall relatively close to a straight line.

Finding the significance of r. You might want to know how significant an r of $-.899$ is. The formula to test the null hypothesis that R (the population correlation) $= 0$ is

$$t = \frac{r\sqrt{(n - 2)}}{\sqrt{(1 - r^2)}}$$

where r is the sample correlation coefficient, and n is the size of the sample (the number of data pairs). The probability of t may be looked up in Table 3 (page 154) using $n - 2$ degrees of freedom.

Example:

$$t = \frac{-.899\sqrt{(10 - 2)}}{\sqrt{1 - (-1.899^2)}} = \frac{-2.543}{.438} = -5.807$$

The probability of obtaining a t of -5.807 with 8 df (drop the sign when looking up the value of t) is lower than the lowest listed probability of .0005. If the correlation between months of exercise-machine ownership and hours of exercise per week were actually 0, you would expect an r of $-.899$ or lower in fewer than one out of a thousand random samples.

To evaluate a correlation coefficient, first determine its significance. If the probability that the coefficient resulted from chance is not acceptably low, the analysis should end there—neither the coefficient's sign nor its magnitude may reflect anything other than sampling error. If the coefficient is statistically significant, the sign should give an indication of the direction of the relationship, and the magnitude indicates its strength. Remember, however, that all statistics become significant with a high enough n.

Even if it's statistically significant, whether a correlation of a given magnitude is substantively or practically significant depends greatly on the phenomenon being studied. Generally, correlations tend to be higher in the physical sciences, where relationships between variables often obey uniform laws, and lower in the social sciences, where relationships may be harder to predict. A correlation of .4 between a pair of sociological variables may be more meaningful than a correlation of .7 between two variables in physics.

Bear in mind also that the correlation coefficient measures only straight-line relationships. Not all relationships between variables trace a straight line. Figure 46 shows a curvilinear relationship such that values of *y* increase along with values of *x* up to a point, then decrease with higher values of *x*. The correlation coefficient for this plot is 0, the same as for the plot in Figure 44. This plot, however, shows a relationship that Figure 44 does not.

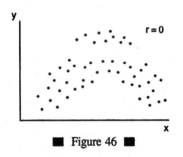

■ Figure 46 ■

Correlation does not imply causation. The fact that Variable A and Variable B are correlated does not necessarily mean that A caused B or that B caused A (though either may be true). If you were to examine a database of demographic information, for example, you would find that the number of churches in a city is correlated with the number of violent crimes in the city. The reason is not that church attendance causes crime, but that these two variables both increase as a function of a third variable: population.

Also note that the scales used to measure two variables have no effect on their correlation. If you had converted hours of exercise per week to minutes per week, and/or months machine owned to days owned, the correlation coefficient would have been the same.

The coefficient of determination. The square of the correlation coefficient *r* is called the **coefficient of determination** and can be used as a measure of the proportion of variability that two variables share, or how much one can be "explained" by the other. The coefficient of determination for this example is $-0.899^2 = .808$. Approximately 80% of the variability of each variable is shared with the other variable.

Simple Linear Regression

Although correlation is not concerned with causation in relationships among variables, a related statistical procedure called **regression** often is. Regression is used to assess the contribution of one or more "causing" variables (called **independent** variables) to one "caused" (or **dependent**) variable. It can also be used to predict the value of one variable from the values of others. When there is only one independent variable and when the relationship can be expressed as a straight line, the procedure is called simple linear regression.

Any straight line in two-dimensional space can be represented by the equation:

$$y = a + bx$$

where y is the variable on the vertical axis, x is the variable on the horizontal axis, a is the y-value where the line crosses the vertical axis (often called the **intercept**), and b is the amount of change in y corresponding to a one-unit increase in x (often called the **slope**). Figure 47 gives an example:

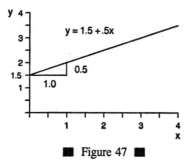

■ Figure 47 ■

Returning to the exercise example, you observed that the scatterplot of points in Figure 42 resembles a line. The regression procedure fits the best possible straight line to an array of data points. If no single line can be drawn such that all the points fall on it, what is the "best" line? The best line is the one that minimizes the distance of all the data points to the line, a fact that will become clearer after you compute the line for the example.

Regression is an inferential procedure, meaning that it can be used to draw conclusions about populations based on samples randomly drawn from those populations. Suppose that your ten exercise-machine owners were randomly selected to represent the population of all exercise-machine owners. In order to use this sample to make educated guesses about the relationship between the two variables (months of machine ownership and time spent exercising) in the population, you will need to rewrite the equation above to reflect the fact that you will be estimating population parameters:

$$y = \beta_0 + \beta_1 x$$

All you have done is replace the intercept (a) with β_0 and the slope (b) with β_1. The formula to compute the parameter estimate $\hat{\beta}_1$ is:

$$\hat{\beta}_1 = \frac{S_{xy}}{S_{xx}}$$

where

$$S_{xy} = \Sigma xy - \frac{(\Sigma x)(\Sigma y)}{n}$$

and

$$S_{xx} = \Sigma x^2 - \frac{(\Sigma x)^2}{n}$$

The formula to compute the parameter estimate $\hat{\beta}_0$ is

$$\hat{\beta}_0 = \bar{y} - \hat{\beta}_1 \bar{x}$$

where $\bar{y}$ and $\bar{x}$ are the two sample means.

You have already computed the quantities that you need to substitute into these formulas for the exercise example—except for the mean of $x(\bar{x})$, which is $64/10 = 6.4$, and the mean of y $(\bar{y})$, which is $56/10 = 5.6$. First, compute the estimate of the slope:

$$S_{xy} = 285 - \frac{(64)(56)}{10} = -73.4$$

$$S_{xx} = 520 - \frac{4096}{10} = 110.4$$

$$\hat{\beta}_1 = \frac{-73.4}{110.4} = -0.665$$

Now the intercept may be computed:

$$\hat{\beta}_0 = \bar{y} - \hat{\beta}_1\bar{x}$$

$$= 5.6 - (-0.665(6.4))$$

$$= 5.6 - (-4.256)$$

$$= 9.856$$

So, the regression equation for the example is $y = 9.856 - 0.665(x)$. When you plot this line over the data points, the result looks like Figure 48.

■ Figure 48 ■

The vertical distance from each data point to the regression line is the error, or **residual**, of the line's accuracy in estimating that point. Some points have positive residuals (they lie above the line), some have negative ones (they lie below it). If all the points fell on the line, there would be no error and no residuals. The mean of the sample residuals is always 0 because the regression line is always drawn such that half of the error is above it and half below. The equations that you used to estimate the intercept and slope determine a line of "best fit" by minimizing the sum of the squared residuals. This method of regression is called **least squares.**

Because regression estimates usually contain some error (that is, all points do not fall on the line), an error term (ε, the Greek letter epsilon) is usually added to the end of the equation:

$$y = \beta_0 + \beta_1 x + \varepsilon$$

The estimate of the slope β_1 for the exercise example was -0.665. The slope is negative because the line slants down from left to right, as it must for two variables that are negatively correlated, reflecting that one variable decreases as the other increases. When the correlation is positive, β_1 is positive, and the line slants up from left to right.

Confidence interval for the slope. What if the slope is 0, as in Figure 49? That means that y has no linear dependence on x, or that knowing x does not contribute anything to your ability to predict y.

■ Figure 49 ■

It is often useful to compute a confidence interval for a regression slope. If it contains 0, you would be unable to conclude that x and y are related. The formula to compute a confidence interval for β_1 is

$$(a,b) = \hat{\beta}_1 \pm t_{\alpha/2, n-2} \cdot \frac{s}{\sqrt{S_{xx}}}$$

where

$$s = \frac{\Sigma(y - \hat{y})^2}{n-2}$$

and

$$S_{xx} = \Sigma x^2 - \frac{(\Sigma x)^2}{n}$$

and where $\Sigma(y - \hat{y})^2$ is the sum of the squared residuals, $t_{\alpha/2, n-2}$ is the critical value from the t-table corresponding to half the desired alpha level at $n - 2$ degrees of freedom, and n is the size of the sample (the number of data pairs). The test for this example will use an alpha of .05. Table 3 (page 154) shows that $t_{.025, 8} = 2.306$.

Compute the quantity $\Sigma(y - \hat{y})^2$ by subtracting each predicted y-value ($\hat{y}$) from each actual y-value, squaring it, and summing the squares. The predicted y-value ($\hat{y}$) is the y-value that would be predicted from each given x using the formula $y = 9.856 - 0.665(x)$.

x	y		$\hat{y}$		residual	residual^2
5	5	−	6.530	=	−1.530	2.341
10	2	−	3.205	=	−1.205	1.452
4	8	−	7.195	=	0.805	.648
8	3	−	4.535	=	−1.535	2.356
2	8	−	8.525	=	−0.525	.276
7	5	−	5.200	=	−0.200	.040
9	5	−	3.870	=	1.130	1.277
6	7	−	5.865	=	1.135	1.288
1	10	−	9.190	=	0.810	.656
12	3	−	1.875	=	1.125	1.266
					0	11.600

Now, compute s:

$$s = \sqrt{\frac{11.600}{10 - 2}} = \sqrt{1.45} = 1.204$$

You have already determined that $S_{xx} = 110.4$; you can proceed to the main formula:

$$(a,b) = -0.665 \pm 2.306 \frac{1.204}{\sqrt{110.4}}$$

$$= -0.665 \pm 2.306 \frac{1.204}{10.507}$$

$$= -0.665 \pm 2.306\,(0.115)$$

$$= -0.665 \pm 0.264$$

$$= (-0.929,\ -0.401)$$

You can be 95% certain that the population parameter β_1 (the slope) is no lower than -0.929 and no higher than -0.401. Because this interval does not contain 0, you would be able to reject the null hypothesis that $\beta_1 = 0$ and conclude that these two variables are indeed related in the population.

Confidence interval for prediction. You have learned that you could predict a y-value from a given x-value. Because there is some error associated with your prediction, however, you might want to produce a confidence interval rather than a simple point estimate. The formula for a prediction interval for y for a given x is

$$(a,b) = \hat{y} \pm t_{\alpha/2,\,n-2}\,(s)\sqrt{1 + \frac{1}{n} + \frac{(x - \bar{x})^2}{S_{xx}}}$$

where

$$S_{xx} = \Sigma x^2 - \frac{(\Sigma x)^2}{n}$$

and where $\hat{y}$ is the y-value predicted for x using the regression equation, $t_{\alpha/2,\,n-2}$ is the critical value from the t-table corresponding to half the desired alpha level at $n - 2$ degrees of freedom, and n is the size of the sample (the number of data pairs).

What is a 90% confidence interval for the number of hours spent exercising per week if the exercise machine is owned 11 months?

The first step is to use the original regression equation to compute a point estimate for y:

$$y = 9.856 - 0.665(11)$$

$$= 9.856 - 7.315$$

$$= 2.541$$

For a 90% confidence interval, you need to use $t_{.05,8}$, which Table 3 (page 154) shows to be 1.860. You have already computed the remaining quantities, so you can proceed with the formula:

$$(a,b) = 2.541 \pm 1.860(1.204)\sqrt{1 + \frac{1}{10} + \frac{(11 - 6.4)^2}{110.40}}$$

$$= 2.541 \pm 1.860(1.204)\sqrt{1 + .10 + \frac{21.16}{110.40}}$$

$$= 2.541 \pm 1.860(1.204)\sqrt{1.292}$$

$$= 2.541 \pm 2.545$$

$$= (-0.004, 5.086)$$

You can be 90% confident that the population mean for the number of hours spent exercising per week when x (number of weeks machine owned) = 11 is between about 0 and 5.

Assumptions and cautions. The use of regression for parametric inference assumes that the errors (ε) are (1) independent of each other and (2) normally distributed with the same variance for each level of the independent variable. Figure 50 shows a violation of (2). The errors (residuals) are greater for higher values of x than for lower values.

■ Figure 50 ■

Least squares regression is sensitive to **outliers,** or data points that fall far from most other points. If you were to add the single data point $x = 15$, $y = 12$ to the exercise data, the regression line would change to the dotted line shown in Figure 51. You need to be wary of outliers because they can influence the regression equation greatly.

■ Figure 51 ■

It can be dangerous to extrapolate in regression—to predict values beyond the range of your dataset. The regression model assumes that the straight line extends to infinity in both directions, which is often not true. According to the regression equation for the example, people who have owned their exercise machines longer than around 15 months do not exercise at all. It is more likely, however, that "hours of exercise"

---

reaches some minimum threshold and then declines only gradually, if at all (Figure 52).

■ Figure 52 ■

Chi-Square

The statistical procedures that you have reviewed thus far are appropriate only for variables at the interval and ratio levels of measurement. The **chi-square** (χ^2) test can be used to evaluate a relationship between two nominal or ordinal variables. It is one example of a **non-parametric test**. Non-parametric tests are used when assumptions about normal distribution in the population cannot be met or when the level of measurement is ordinal or less. These tests are less powerful than parametric tests.

Suppose that 125 children are shown three television commercials for breakfast cereal and are asked to pick which they liked best. The results are shown in the table on the following page.

You would like to know if the choice of favorite commercial was related to whether the child was a boy or a girl or if these two variables are independent. The totals in the margins will allow you to determine the overall probability of (1) liking commercial A, B, or C, regardless of gender, and (2) being either a boy or a girl, regardless of favorite commercial. If the two variables are independent, then you should be

FAVORITE COMMERCIAL

	A	B	C	Totals
Boys	30	29	16	75
Girls	12	33	5	50
Totals	42	62	21	125

able to use these probabilities to predict approximately how many children should be in each cell. If the actual count is very different from the count that you would expect if the probabilities are independent, the two variables must be related.

Consider the first (upper left) cell of the table. The overall probability of a child in the sample being a boy is $75 \div 125 = .6$. The overall probability of liking Commercial A is $42 \div 125 = .336$. The multiplication rule that you learned in the section on probability (page 41) states that the probability of both of two independent events occurring is the product of their two probabilities. Therefore, the probability of a child both being a boy and liking Commercial A is $.6 \times .336 = .202$. The expected number of children in this cell, then, is $.202 \times 125 = 25.2$.

There is a faster way of computing the expected count for each cell: multiply the row total by the column total and divide by n. The expected count for the first cell is, therefore, $(75 \times 42) \div 125 = 25.2$. If you perform this operation for each cell, you get the following expected counts (in parentheses):

	FAVORITE COMMERCIAL			
	A	**B**	**C**	Totals
Boys	30 (25.2)	29 (37.2)	16 (12.6)	75
Girls	12 (16.8)	33 (24.8)	5 (8.4)	50
Totals	42	62	21	125

Note that the expected counts properly add up to the row and column totals. You are now ready for the formula for χ^2, which compares each cell's actual count to its expected count:

$$\chi^2 = \Sigma \frac{(\text{observed} - \text{expected})^2}{\text{expected}}$$

The formula describes an operation that is performed on each cell and which yields a number. When all the numbers are summed, the result is χ^2. Now, compute it for the six cells in the example:

$$\chi^2 = \frac{(30 - 25.2)^2}{25.2} + \frac{(29 - 37.2)^2}{37.2} + \frac{(16 - 12.6)^2}{12.6} +$$

$$\frac{(12 - 16.8)^2}{16.8} + \frac{(33 - 24.8)^2}{24.8} + \frac{(5 - 8.4)^2}{8.4}$$

$$= .914 + 1.808 + .917 + 1.371 + 2.711 + 1.376 = 9.097$$

The larger χ^2, the more likely that the variables are related; note that the cells that contribute the most to the resulting statistic are those in which the expected count is very different from the actual count.

Chi-square has a probability distribution, the critical values for which are listed in Table 4 (page 156). As with the t-distribution, χ^2 has a degrees-of-freedom parameter, the formula for which is

$$(\# \text{rows} - 1) \times (\# \text{columns} - 1)$$

or in your example:

$$(2 - 1) \times (3 - 1) = 1 \times 2 = 2$$

In Table 4, a chi-square of 9.097 with two degrees of freedom falls between the commonly used significance levels of .05 and .01. If you had specified an alpha of .05 for the test, you could, therefore, reject the null hypothesis that gender and favorite commercial are independent. At $\alpha = .01$, however, you could not reject the null hypothesis.

The χ^2 test does not allow you to conclude anything more specific than that there is some relationship in your sample between gender and commercial liked (at $\alpha = .05$). Examining the observed vs. expected counts in each cell might give you a clue as to the nature of the relationship and which levels of the variables are involved. For example, Commercial B appears to have been liked more by girls than boys. But χ^2 tests only the very general null hypothesis that the two variables are independent.

It may seem that there are many ways to make errors in working a statistics problem. In fact, most errors on statistics exams can be reduced to a short list of common oversights. If you learn to avoid the mistakes listed below, you can greatly reduce your chances of making an error on an exam.

- **Forgetting to convert between standard deviation (σ and s) and variance (σ^2 and s^2).** Some formulas use one; some use the other. Square the standard deviation to get the variance, or take the positive square root of the variance to get the standard deviation.
- **Misstating one-tailed and two-tailed hypotheses.** If the hypothesis predicts simply that one value will be higher than another, it requires a one-tailed test. If, however, it predicts that two values will be different—that is, one value will be either higher *or* lower than another or that they will be equal—then use a two-tailed test. Make sure your null and alternative hypotheses together cover all possibilities—greater than, less than, and equal to.
- **Failing to split the alpha level for two-tailed tests.** If the overall significance level for the test is .05, then you must look up the critical (tabled) value for a probability of .025. The alpha level is always split when computing confidence intervals.
- **Misreading the standard normal (z) table.** As explained in Sampling (page 55), all standard normal tables do not have the same format, and it is important to know what area of the curve (or probability) the table presents as corresponding to a given z-score. Table 2 (page 150) gives the area of the curve lying at or below z. The area to the right of z (or the probability of obtaining a value above z) is simply 1 minus the tabled probability.
- **Using n instead of $n - 1$ degrees of freedom in one-sample t-tests.** Remember that you must subtract 1 from n in order

to get the degrees-of-freedom parameter that you need in order to look up a value in the *t*-table.

- **Confusing confidence level with confidence interval.** The **confidence level** is the significance level of the test or the likelihood of obtaining a given result by chance. The **confidence interval** is a range of values between the lowest and highest values that the estimated parameter could take at a given confidence level.

- **Confusing interval width with margin of error.** A confidence interval is always a point estimate plus or minus a margin of error. The interval width is double that margin of error. If, for example, a population parameter is estimated to be 46% plus or minus 4%, the interval width is 8%.

- **Confusing statistics with parameters.** Parameters are characteristics of the population that you usually do not know; they are designated with Greek symbols (μ and σ). Statistics are characteristics of samples that you are usually able to compute. Although statistics correspond to parameters ($\bar{x}$ is the mean of a sample, as μ is the mean of a population), the two are not interchangeable; hence, you need to be careful and know which variables are parameters and which are statistics. You compute statistics in order to estimate parameters.

- **Confusing the addition rule with the multiplication rule.** When determining probability, the multiplication rule applies if *all* favorable outcomes must occur in a series of events. The addition rule applies when *at least one* success must occur in a series of events.

- **Forgetting that the addition rule applies to mutually exclusive outcomes.** If outcomes can occur together, the probability of their joint occurrence must be subtracted from the total "addition-rule" probability. (See Probability, page 41.)

- **Forgetting to average the two middle values of an even-numbered set when figuring median and (sometimes) quartiles.** If an ordered series contains an even number of measures, the median is always the mean of the two middle measures.

- **Forgetting where to place the points of a frequency polygon.**
  The points of a frequency polygon are always at the center of
  each of the class intervals, not at the ends.

TABLES

TABLE 1
Binomial Probabilities, $P(x)$ for $n \leq 20$

n=2

$x\downarrow$	0.05	0.10	0.15	0.20	π 0.25	0.30	0.35	0.40	0.45	0.50	$x\uparrow$
0	.9025	.8100	.7225	.6400	.5625	.4900	.4225	.3600	.3025	.2500	2
1	.0950	.1800	.2550	.3200	.3750	.4200	.4550	.4800	.4950	.5000	1
2	.0025	.0100	.0225	.0400	.0625	.0900	.1225	.1600	.2025	.2500	0
	0.95	0.90	0.85	0.80	0.75	0.70	0.65	0.60	0.55	0.50	$x\uparrow$

n=3

$x\downarrow$	0.05	0.10	0.15	0.20	π 0.25	0.30	0.35	0.40	0.45	0.50	$x\uparrow$
0	.8574	.7290	.6141	.5120	.4219	.3430	.2746	.2160	.1664	.1250	3
1	.1354	.2430	.3251	.3840	.4219	.4410	.4436	.4320	.4084	.3750	2
2	.0071	.0270	.0574	.0960	.1406	.1890	.2389	.2880	.3341	.3750	1
3	.0001	.0010	.0034	.0080	.0156	.0270	.0429	.0640	.0911	.1250	0
	0.95	0.90	0.85	0.80	0.75	0.70	0.65	0.60	0.55	0.50	$x\uparrow$

TABLE 1

n=4

x↓	0.05	0.10	0.15	0.20	π 0.25	0.30	0.35	0.40	0.45	0.50	x↑
0	.8145	.6561	.5220	.4096	.3164	.2401	.1785	.1296	.0915	.0625	4
1	.1715	.2916	.3685	.4096	.4219	.4116	.3845	.3456	.2995	.2500	3
2	.0135	.0486	.0975	.1536	.2109	.2646	.3105	.3456	.3675	.3750	2
3	.0005	.0036	.0115	.0256	.0469	.0756	.1115	.1536	.2005	.2500	1
4	.0000	.0001	.0005	.0016	.0039	.0081	.0150	.0256	.0410	.0625	0
	0.95	0.90	0.85	0.80	0.75	0.70	0.65	0.60	0.55	0.50	x↑

n=5

x↓	0.05	0.10	0.15	0.20	π 0.25	0.30	0.35	0.40	0.45	0.50	x↑
0	.7738	.5905	.4437	.3277	.2373	.1681	.1160	.0778	.0503	.0313	5
1	.2036	.3281	.3915	.4096	.3955	.3602	.3124	.2592	.2059	.1563	4
2	.0214	.0729	.1382	.2048	.2637	.3087	.3364	.3456	.3369	.3125	3
3	.0011	.0081	.0244	.0512	.0879	.1323	.1811	.2304	.2757	.3125	2
4	.0000	.0005	.0022	.0064	.0146	.0284	.0488	.0768	.1128	.1563	1
5	.0000	.0000	.0001	.0003	.0010	.0024	.0053	.0102	.0185	.0313	0
	0.95	0.90	0.85	0.80	0.75	0.70	0.65	0.60	0.55	0.50	x↑

TABLE 1

n=6

x↓	0.05	0.10	0.15	0.20	π 0.25	0.30	0.35	0.40	0.45	0.50	x↑
0	.7351	.5314	.3771	.2621	.1780	.1176	.0754	.0467	.0277	.0156	6
1	.2321	.3543	.3993	.3932	.3560	.3025	.2437	.1866	.1359	.0938	5
2	.0305	.0984	.1762	.2458	.2966	.3241	.3280	.3110	.2780	.2344	4
3	.0021	.0146	.0415	.0819	.1318	.1852	.2355	.2765	.3032	.3125	3
4	.0001	.0012	.0055	.0154	.0330	.0595	.0951	.1382	.1861	.2344	2
5	.0000	.0001	.0004	.0015	.0044	.0102	.0205	.0369	.0609	.0938	1
6	.0000	.0000	.0000	.0001	.0002	.0007	.0018	.0041	.0083	.0156	0
	0.95	0.90	0.85	0.80	0.75	0.70	0.65	0.60	0.55	0.50	

TABLE 1

n=7

x↓	0.05	0.10	0.15	0.20	π 0.25	0.30	0.35	0.40	0.45	0.50	x↑
0	.6983	.4783	.3206	.2097	.1335	.0824	.0490	.0280	.0152	.0078	7
1	.2573	.3720	.3960	.3670	.3115	.2471	.1848	.1306	.0872	.0547	6
2	.0406	.1240	.2097	.2753	.3115	.3177	.2985	.2613	.2140	.1641	5
3	.0036	.0230	.0617	.1147	.1730	.2269	.2679	.2903	.2918	.2734	4
4	.0002	.0026	.0109	.0287	.0577	.0972	.1442	.1935	.2388	.2734	3
5	.0000	.0002	.0012	.0043	.0115	.0250	.0466	.0774	.1172	.1641	2
6	.0000	.0000	.0001	.0004	.0013	.0036	.0084	.0172	.0320	.0547	1
7	.0000	.0000	.0000	.0000	.0001	.0002	.0006	.0016	.0037	.0078	0
	0.95	0.90	0.85	0.80	0.75	0.70	0.65	0.60	0.55	0.50	

TABLE 1

n=8

x↓	0.05	0.10	0.15	0.20	π 0.25	0.30	0.35	0.40	0.45	0.50	x↑
0	.6634	.4305	.2725	.1678	.1001	.0576	.0319	.0168	.0084	.0039	8
1	.2793	.3826	.3847	.3355	.2670	.1977	.1373	.0896	.0548	.0313	7
2	.0515	.1488	.2376	.2936	.3115	.2965	.2587	.2090	.1569	.1094	6
3	.0054	.0331	.0839	.1468	.2076	.2541	.2786	.2787	.2568	.2188	5
4	.0004	.0046	.0185	.0459	.0865	.1361	.1875	.2322	.2627	.2734	4
5	.0000	.0004	.0026	.0092	.0231	.0467	.0808	.1239	.1719	.2188	3
6	.0000	.0000	.0002	.0011	.0038	.0100	.0217	.0413	.0703	.1094	2
7	.0000	.0000	.0000	.0001	.0004	.0012	.0033	.0079	.0164	.0313	1
8	.0000	.0000	.0000	.0000	.0000	.0001	.0002	.0007	.0017	.0039	0
	0.95	0.90	0.85	0.80	0.75	0.70	0.65	0.60	0.55	0.50	

TABLE 1

n = 9

$x\downarrow$	0.05	0.10	0.15	0.20	0.25 π	0.30	0.35	0.40	0.45	0.50	$x\uparrow$
0	.6302	.3874	.2316	.1342	.0751	.0404	.0207	.0101	.0046	.0020	9
1	.2985	.3874	.3679	.3020	.2253	.1556	.1004	.0605	.0339	.0176	8
2	.0629	.1722	.2597	.3020	.3003	.2668	.2162	.1612	.1110	.0703	7
3	.0077	.0446	.1069	.1762	.2336	.2668	.2716	.2508	.2119	.1641	6
4	.0006	.0074	.0283	.0661	.1168	.1715	.2194	.2508	.2600	.2461	5
5	.0000	.0008	.0050	.0165	.0389	.0735	.1181	.1672	.2128	.2461	4
6	.0000	.0001	.0006	.0028	.0087	.0210	.0424	.0743	.1160	.1641	3
7	.0000	.0000	.0000	.0003	.0012	.0039	.0098	.0212	.0407	.0703	2
8	.0000	.0000	.0000	.0000	.0001	.0004	.0013	.0035	.0083	.0176	1
9	.0000	.0000	.0000	.0000	.0000	.0000	.0001	.0003	.0008	.0020	0
	0.95	0.90	0.85	0.80	0.75	0.70	0.65	0.60	0.55	0.50	

TABLE 1

n=10

$x\downarrow$	0.05	0.10	0.15	0.20	π 0.25	0.30	0.35	0.40	0.45	0.50	$x\uparrow$
0	.5987	.3487	.1969	.1074	.0563	.0282	.0135	.0060	.0025	.0010	10
1	.3151	.3874	.3474	.2684	.1877	.1211	.0725	.0403	.0207	.0098	9
2	.0746	.1937	.2759	.3020	.2816	.2335	.1757	.1209	.0763	.0439	8
3	.0105	.0574	.1298	.2013	.2503	.2668	.2522	.2150	.1665	.1172	7
4	.0010	.0112	.0401	.0881	.1460	.2001	.2377	.2508	.2384	.2051	6
5	.0001	.0015	.0085	.0264	.0584	.1029	.1536	.2007	.2340	.2461	5
6	.0000	.0001	.0012	.0055	.0162	.0368	.0689	.1115	.1596	.2051	4
7	.0000	.0000	.0001	.0008	.0031	.0090	.0212	.0425	.0746	.1172	3
8	.0000	.0000	.0000	.0001	.0004	.0014	.0043	.0106	.0229	.0439	2
9	.0000	.0000	.0000	.0000	.0000	.0001	.0005	.0016	.0042	.0098	1
10	.0000	.0000	.0000	.0000	.0000	.0000	.0000	.0001	.0003	.0010	0
	0.95	0.90	0.85	0.80	0.75	0.70	0.65	0.60	0.55	0.50	

n=12						π					
x↓	0.05	0.10	0.15	0.20	0.25	0.30	0.35	0.40	0.45	0.50	
0	.5404	.2824	.1422	.0687	.0317	.0138	.0057	.0022	.0008	.0002	12
1	.3413	.3766	.3012	.2062	.1267	.0712	.0368	.0174	.0075	.0029	11
2	.0988	.2301	.2924	.2835	.2323	.1678	.1088	.0639	.0339	.0161	10
3	.0173	.0852	.1720	.2362	.2581	.2397	.1954	.1419	.0923	.0537	9
4	.0021	.0213	.0683	.1329	.1936	.2311	.2367	.2128	.1700	.1208	8
5	.0002	.0038	.0193	.0532	.1032	.1585	.2039	.2270	.2225	.1934	7
6	.0000	.0005	.0040	.0155	.0401	.0792	.1281	.1766	.2124	.2256	6
7	.0000	.0000	.0006	.0033	.0115	.0291	.0591	.1009	.1489	.1934	5
8	.0000	.0000	.0001	.0005	.0024	.0078	.0199	.0420	.0762	.1208	4
9	.0000	.0000	.0000	.0001	.0004	.0015	.0048	.0125	.0277	.0537	3
10	.0000	.0000	.0000	.0000	.0000	.0002	.0008	.0025	.0068	.0161	2
11	.0000	.0000	.0000	.0000	.0000	.0000	.0001	.0003	.0010	.0029	1
12	.0000	.0000	.0000	.0000	.0000	.0000	.0000	.0000	.0001	.0002	0
	0.95	0.90	0.85	0.80	0.75	0.70	0.65	0.60	0.55	0.50	x↑

TABLE 1

TABLE 1

$n=14$						π					
$x\downarrow$	0.05	0.10	0.15	0.20	0.25	0.30	0.35	0.40	0.45	0.50	$x\uparrow$
0	.4877	.2288	.1028	.0440	.0178	.0068	.0024	.0008	.0002	.0001	14
1	.3593	.3559	.2539	.1539	.0832	.0407	.0181	.0073	.0027	.0009	13
2	.1229	.2570	.2912	.2501	.1802	.1134	.0634	.0317	.0141	.0056	12
3	.0259	.1142	.2056	.2501	.2402	.1943	.1366	.0845	.0462	.0222	11
4	.0037	.0349	.0998	.1720	.2202	.2290	.2022	.1549	.1040	.0611	10
5	.0004	.0078	.0352	.0860	.1468	.1963	.2178	.2066	.1701	.1222	9
6	.0000	.0013	.0093	.0322	.0734	.1262	.1759	.2066	.2088	.1833	8
7	.0000	.0002	.0019	.0092	.0280	.0618	.1082	.1574	.1952	.2095	7
8	.0000	.0000	.0003	.0020	.0082	.0232	.0510	.0918	.1398	.1833	6
9	.0000	.0000	.0000	.0003	.0018	.0066	.0183	.0408	.0762	.1222	5
10	.0000	.0000	.0000	.0000	.0003	.0014	.0049	.0136	.0312	.0611	4
11	.0000	.0000	.0000	.0000	.0000	.0002	.0010	.0033	.0093	.0222	3
12	.0000	.0000	.0000	.0000	.0000	.0000	.0001	.0005	.0019	.0056	2
13	.0000	.0000	.0000	.0000	.0000	.0000	.0000	.0001	.0002	.0009	1
14	.0000	.0000	.0000	.0000	.0000	.0000	.0000	.0000	.0000	.0001	0
	0.95	0.90	0.85	0.80	0.75	0.70	0.65	0.60	0.55	0.50	

TABLE 1

n=16						π					
x↓	0.05	0.10	0.15	0.20	0.25	0.30	0.35	0.40	0.45	0.50	x↑
0	.4401	.1853	.0743	.0281	.0100	.0033	.0010	.0003	.0001	.0000	16
1	.3706	.3294	.2097	.1126	.0535	.0228	.0087	.0030	.0009	.0002	15
2	.1463	.2745	.2775	.2111	.1336	.0732	.0353	.0150	.0056	.0018	14
3	.0359	.1423	.2285	.2463	.2079	.1465	.0888	.0468	.0215	.0085	13
4	.0061	.0514	.1311	.2001	.2252	.2040	.1553	.1014	.0572	.0278	12
5	.0008	.0137	.0555	.1201	.1802	.2099	.2008	.1623	.1123	.0667	11
6	.0001	.0028	.0180	.0550	.1101	.1649	.1982	.1983	.1684	.1222	10
7	.0000	.0004	.0045	.0197	.0524	.1010	.1524	.1889	.1969	.1746	9
8	.0000	.0001	.0009	.0055	.0197	.0487	.0923	.1417	.1812	.1964	8
9	.0000	.0000	.0001	.0012	.0058	.0185	.0442	.0840	.1318	.1746	7
10	.0000	.0000	.0000	.0002	.0014	.0056	.0167	.0392	.0755	.1222	6
11	.0000	.0000	.0000	.0000	.0002	.0013	.0049	.0142	.0337	.0667	5
12	.0000	.0000	.0000	.0000	.0000	.0002	.0011	.0040	.0115	.0278	4
13	.0000	.0000	.0000	.0000	.0000	.0000	.0002	.0008	.0029	.0085	3
14	.0000	.0000	.0000	.0000	.0000	.0000	.0000	.0001	.0005	.0018	2
15	.0000	.0000	.0000	.0000	.0000	.0000	.0000	.0000	.0001	.0002	1
	0.95	0.90	0.85	0.80	0.75	0.70	0.65	0.60	0.55	0.50	x↑

TABLE 1

n=18

x↓	0.05	0.10	0.15	0.20	π 0.25	0.30	0.35	0.40	0.45	0.50	x↑
0	.3972	.1501	.0536	.0180	.0056	.0016	.0004	.0001	.0000	.0000	18
1	.3763	.3002	.1704	.0811	.0338	.0126	.0042	.0012	.0003	.0001	17
2	.1683	.2835	.2556	.1723	.0958	.0458	.0190	.0069	.0022	.0006	16
3	.0473	.1680	.2406	.2297	.1704	.1046	.0547	.0246	.0095	.0031	15
4	.0093	.0700	.1592	.2153	.2130	.1681	.1104	.0614	.0291	.0117	14
5	.0014	.0218	.0787	.1507	.1988	.2017	.1664	.1146	.0666	.0327	13
6	.0002	.0052	.0301	.0816	.1436	.1873	.1941	.1655	.1181	.0708	12
7	.0000	.0010	.0091	.0350	.0820	.1376	.1792	.1892	.1657	.1214	11
8	.0000	.0002	.0022	.0120	.0376	.0811	.1327	.1734	.1864	.1669	10
9	.0000	.0000	.0004	.0033	.0139	.0386	.0794	.1284	.1694	.1855	9
10	.0000	.0000	.0001	.0008	.0042	.0149	.0385	.0771	.1248	.1669	8
11	.0000	.0000	.0000	.0001	.0010	.0046	.0151	.0374	.0742	.1214	7
12	.0000	.0000	.0000	.0000	.0002	.0012	.0047	.0145	.0354	.0708	6
13	.0000	.0000	.0000	.0000	.0000	.0002	.0012	.0045	.0134	.0327	5
14	.0000	.0000	.0000	.0000	.0000	.0000	.0002	.0011	.0039	.0117	4
15	.0000	.0000	.0000	.0000	.0000	.0000	.0000	.0002	.0009	.0031	3
16	.0000	.0000	.0000	.0000	.0000	.0000	.0000	.0000	.0001	.0006	2
17	.0000	.0000	.0000	.0000	.0000	.0000	.0000	.0000	.0000	.0001	1
	0.95	0.90	0.85	0.80	0.75	0.70	0.65	0.60	0.55	0.50	

TABLE 1

$n=20$						π					
$x\downarrow$	0.05	0.10	0.15	0.20	0.25	0.30	0.35	0.40	0.45	0.50	$x\uparrow$
0	.3585	.1216	.0388	.0115	.0032	.0008	.0002	.0000	.0000	.0000	20
1	.3774	.2702	.1368	.0576	.0211	.0068	.0020	.0005	.0001	.0000	19
2	.1887	.2852	.2293	.1369	.0669	.0278	.0100	.0031	.0008	.0002	18
3	.0596	.1901	.2428	.2054	.1339	.0716	.0323	.0123	.0040	.0011	17
4	.0133	.0898	.1821	.2182	.1897	.1304	.0738	.0350	.0139	.0046	16
5	.0022	.0319	.1028	.1746	.2023	.1789	.1272	.0746	.0365	.0148	15
6	.0003	.0089	.0454	.1091	.1686	.1916	.1712	.1244	.0746	.0370	14
7	.0000	.0020	.0160	.0545	.1124	.1643	.1844	.1659	.1221	.0739	13
8	.0000	.0004	.0046	.0222	.0609	.1144	.1614	.1797	.1623	.1201	12
9	.0000	.0001	.0011	.0074	.0271	.0654	.1158	.1597	.1771	.1602	11
10	.0000	.0000	.0002	.0020	.0099	.0308	.0686	.1171	.1593	.1762	10
11	.0000	.0000	.0000	.0005	.0030	.0120	.0336	.0710	.1185	.1602	9
12	.0000	.0000	.0000	.0001	.0008	.0039	.0136	.0355	.0727	.1201	8
13	.0000	.0000	.0000	.0000	.0002	.0010	.0045	.0146	.0366	.0739	7
14	.0000	.0000	.0000	.0000	.0000	.0002	.0012	.0049	.0150	.0370	6
15	.0000	.0000	.0000	.0000	.0000	.0000	.0003	.0013	.0049	.0148	5
16	.0000	.0000	.0000	.0000	.0000	.0000	.0000	.0003	.0013	.0046	4
17	.0000	.0000	.0000	.0000	.0000	.0000	.0000	.0000	.0002	.0011	3
18	.0000	.0000	.0000	.0000	.0000	.0000	.0000	.0000	.0000	.0002	2
	0.95	0.90	0.85	0.80	0.75	0.70	0.65	0.60	0.55	0.50	

TABLE 2

TABLE 2

Standard Normal Probabilities

(Table entry is probability at or below z.)

Probability

z

z	.00	.01	.02	.03	.04
-3.4	.0003	.0003	.0003	.0003	.0003
-3.3	.0005	.0005	.0005	.0004	.0004
-3.2	.0007	.0007	.0006	.0006	.0006
-3.1	.0010	.0009	.0009	.0009	.0008
-3.0	.0013	.0013	.0013	.0012	.0012
-2.9	.0019	.0018	.0018	.0017	.0016
-2.8	.0026	.0025	.0024	.0023	.0023
-2.7	.0035	.0034	.0033	.0032	.0031
-2.6	.0047	.0045	.0044	.0043	.0041
-2.5	.0062	.0060	.0059	.0057	.0055
-2.4	.0082	.0080	.0078	.0075	.0073
-2.3	.0107	.0104	.0102	.0099	.0096
-2.2	.0139	.0136	.0132	.0129	.0125
-2.1	.0179	.0174	.0170	.0166	.0162
-2.0	.0228	.0222	.0217	.0212	.0207
-1.9	.0287	.0281	.0274	.0268	.0262
-1.8	.0359	.0351	.0344	.0336	.0329
-1.7	.0446	.0436	.0427	.0418	.0409
-1.6	.0548	.0537	.0526	.0516	.0505
-1.5	.0668	.0655	.0643	.0630	.0618
-1.4	.0808	.0793	.0778	.0764	.0749
-1.3	.0968	.0951	.0934	.0918	.0901
-1.2	.1151	.1131	.1112	.1093	.1075
-1.1	.1357	.1335	.1314	.1292	.1271
-1.0	.1587	.1562	.1539	.1515	.1492
-0.9	.1841	.1814	.1788	.1762	.1736
-0.8	.2119	.2090	.2061	.2033	.2005
-0.7	.2420	.2389	.2358	.2327	.2296
-0.6	.2743	.2709	.2676	.2643	.2611
-0.5	.3085	.3050	.3015	.2981	.2946
-0.4	.3446	.3409	.3372	.3336	.3300
-0.3	.3821	.3783	.3745	.3707	.3669
-0.2	.4207	.4168	.4129	.4090	.4052
-0.1	.4602	.4562	.4522	.4483	.4443
-0.0	.5000	.4960	.4920	.4880	.4840

TABLE 2

.05	.06	.07	.08	.09	z
.0003	.0003	.0003	.0003	.0002	−3.4
.0004	.0004	.0004	.0004	.0003	−3.3
.0006	.0006	.0005	.0005	.0005	−3.2
.0008	.0008	.0008	.0007	.0007	−3.1
.0011	.0011	.0011	.0010	.0010	−3.0
.0016	.0015	.0015	.0014	.0014	−2.9
.0022	.0021	.0021	.0020	.0019	−2.8
.0030	.0029	.0028	.0027	.0026	−2.7
.0040	.0039	.0038	.0037	.0036	−2.6
.0054	.0052	.0051	.0049	.0048	−2.5
.0071	.0069	.0068	.0066	.0064	−2.4
.0094	.0091	.0089	.0087	.0084	−2.3
.0122	.0119	.0116	.0113	.0110	−2.2
.0158	.0154	.0150	.0146	.0143	−2.1
.0202	.0197	.0192	.0188	.0183	−2.0
.0256	.0250	.0244	.0239	.0233	−1.9
.0322	.0314	.0307	.0301	.0294	−1.8
.0401	.0392	.0384	.0375	.0367	−1.7
.0495	.0485	.0475	.0465	.0455	−1.6
.0606	.0594	.0582	.0571	.0559	−1.5
.0735	.0721	.0708	.0694	.0681	−1.4
.0885	.0869	.0853	.0838	.0823	−1.3
.1056	.1038	.1020	.1003	.0985	−1.2
.1251	.1230	.1210	.1190	.1170	−1.1
.1469	.1446	.1423	.1401	.1379	−1.0
.1711	.1685	.1660	.1635	.1611	−0.9
.1977	.1949	.1922	.1894	.1867	−0.8
.2266	.2236	.2206	.2177	.2148	−0.7
.2578	.2546	.2514	.2483	.2451	−0.6
.2912	.2877	.2843	.2810	.2776	−0.5
.3264	.3228	.3192	.3156	.3121	−0.4
.3632	.3594	.3557	.3520	.3483	−0.3
.4013	.3974	.3936	.3897	.3859	−0.2
.4404	.4364	.4325	.4286	.4247	−0.1
.4801	.4761	.4721	.4681	.4641	−0.0

TABLE 2

TABLE 2 (*Continued*)
Standard Normal Probabilities
(Table entry is probability at or below *z*.)

z	.00	.01	.02	.03	.04
0.0	.5000	.5040	.5080	.5120	.5160
0.1	.5398	.5438	.5478	.5517	.5557
0.2	.5793	.5832	.5871	.5910	.5948
0.3	.6179	.6217	.6255	.6293	.6331
0.4	.6554	.6591	.6628	.6664	.6700
0.5	.6915	.6950	.6985	.7019	.7054
0.6	.7257	.7291	.7324	.7357	.7389
0.7	.7580	.7611	.7642	.7673	.7704
0.8	.7881	.7910	.7939	.7967	.7995
0.9	.8159	.8186	.8212	.8238	.8264
1.0	.8413	.8438	.8461	.8485	.8508
1.1	.8643	.8665	.8686	.8708	.8729
1.2	.8849	.8869	.8888	.8907	.8925
1.3	.9032	.9049	.9066	.9082	.9099
1.4	.9192	.9207	.9222	.9236	.9251
1.5	.9332	.9345	.9357	.9370	.9382
1.6	.9452	.9463	.9474	.9484	.9495
1.7	.9554	.9564	.9573	.9582	.9591
1.8	.9641	.9649	.9656	.9664	.9671
1.9	.9713	.9719	.9726	.9732	.9738
2.0	.9772	.9778	.9783	.9788	.9793
2.1	.9821	.9826	.9830	.9834	.9838
2.2	.9861	.9864	.9868	.9871	.9875
2.3	.9893	.9896	.9898	.9901	.9904
2.4	.9918	.9920	.9922	.9925	.9927
2.5	.9938	.9940	.9941	.9943	.9945
2.6	.9953	.9955	.9956	.9957	.9959
2.7	.9965	.9966	.9967	.9968	.9969
2.8	.9974	.9975	.9976	.9977	.9977
2.9	.9981	.9982	.9982	.9983	.9984
3.0	.9987	.9987	.9987	.9988	.9988
3.1	.9990	.9991	.9991	.9991	.9992
3.2	.9993	.9993	.9994	.9994	.9994
3.3	.9995	.9995	.9995	.9996	.9996
3.4	.9997	.9997	.9997	.9997	.9997

TABLE 2

.05	.06	.07	.08	.09	z
.5199	.5239	.5279	.5319	.5359	0.0
.5596	.5636	.5675	.5714	.5753	0.1
.5987	.6026	.6064	.6103	.6141	0.2
.6368	.6406	.6443	.6480	.6517	0.3
.6736	.6772	.6808	.6844	.6879	0.4
.7088	.7123	.7157	.7190	.7224	0.5
.7422	.7454	.7486	.7517	.7549	0.6
.7734	.7764	.7794	.7823	.7852	0.7
.8023	.8051	.8078	.8106	.8133	0.8
.8289	.8315	.8340	.8365	.8389	0.9
.8531	.8554	.8577	.8599	.8621	1.0
.8749	.8770	.8790	.8810	.8830	1.1
.8944	.8962	.8980	.8997	.9015	1.2
.9115	.9131	.9147	.9162	.9177	1.3
.9265	.9279	.9292	.9306	.9319	1.4
.9394	.9406	.9418	.9429	.9441	1.5
.9505	.9515	.9525	.9535	.9545	1.6
.9599	.9608	.9616	.9625	.9633	1.7
.9678	.9686	.9693	.9699	.9706	1.8
.9744	.9750	.9756	.9761	.9767	1.9
.9798	.9803	.9808	.9812	.9817	2.0
.9842	.9846	.9850	.9854	.9857	2.1
.9878	.9881	.9884	.9887	.9890	2.2
.9906	.9909	.9911	.9913	.9916	2.3
.9929	.9931	.9932	.9934	.9936	2.4
.9946	.9948	.9949	.9951	.9952	2.5
.9960	.9961	.9962	.9963	.9964	2.6
.9970	.9971	.9972	.9973	.9974	2.7
.9978	.9979	.9979	.9980	.9981	2.8
.9984	.9985	.9985	.9986	.9986	2.9
.9989	.9989	.9989	.9990	.9990	3.0
.9992	.9992	.9992	.9993	.9993	3.1
.9994	.9994	.9995	.9995	.9995	3.2
.9996	.9996	.9996	.9996	.9997	3.3
.9997	.9997	.9997	.9997	.9998	3.4

TABLE 3

TABLE 3
t Distribution Critical Values
(Table entry is the point $t*$ with given probability p lying above it.)

Probability
$t*$

df	.25	.20	.15	.10	.05	.025	.02	.01	.005	.0025	.001	.0005
1	1.000	1.376	1.963	3.078	6.314	12.71	15.89	31.82	63.66	127.3	318.3	636.6
2	.816	1.061	1.386	1.886	2.920	4.303	4.849	6.965	9.925	14.09	22.33	31.60
3	.765	.978	1.250	1.638	2.353	3.182	3.482	4.541	5.841	7.453	10.21	12.92
4	.741	.941	1.190	1.533	2.132	2.776	2.999	3.747	4.604	5.598	7.173	8.610
5	.727	.920	1.156	1.476	2.015	2.571	2.757	3.365	4.032	4.773	5.893	6.869
6	.718	.906	1.134	1.440	1.943	2.447	2.612	3.143	3.707	4.317	5.208	5.959
7	.711	.896	1.119	1.415	1.895	2.365	2.517	2.998	3.499	4.029	4.785	5.408
8	.706	.889	1.108	1.397	1.860	2.306	2.449	2.896	3.355	3.833	4.501	5.041
9	.703	.883	1.100	1.383	1.833	2.262	2.398	2.821	3.250	3.690	4.297	4.781
10	.700	.879	1.093	1.372	1.812	2.228	2.359	2.764	3.169	3.581	4.144	4.587
11	.697	.876	1.088	1.363	1.796	2.201	2.328	2.718	3.106	3.497	4.025	4.437
12	.695	.873	1.083	1.356	1.782	2.179	2.303	2.681	3.055	3.428	3.930	4.318
13	.694	.870	1.079	1.350	1.771	2.160	2.282	2.650	3.012	3.372	3.852	4.221
14	.692	.868	1.076	1.345	1.761	2.145	2.264	2.624	2.977	3.326	3.787	4.140
15	.691	.866	1.074	1.341	1.753	2.131	2.249	2.602	2.947	3.286	3.733	4.073
16	.690	.865	1.071	1.337	1.746	2.120	2.235	2.583	2.921	3.252	3.686	4.015

p

TABLE 3

df						p						
	.25	.20	.15	.10	.05	.025	.02	.01	.005	.0025	.001	.0005
17	.689	.863	1.069	1.333	1.740	2.110	2.224	2.567	2.898	3.222	3.646	3.965
18	.688	.862	1.067	1.330	1.734	2.101	2.214	2.552	2.878	3.197	3.611	3.922
19	.688	.861	1.066	1.328	1.729	2.093	2.205	2.539	2.861	3.174	3.579	3.883
20	.687	.860	1.064	1.325	1.725	2.086	2.197	2.528	2.845	3.153	3.552	3.850
21	.686	.859	1.063	1.323	1.721	2.080	2.189	2.518	2.831	3.135	3.527	3.819
22	.686	.858	1.061	1.321	1.717	2.074	2.183	2.508	2.819	3.119	3.505	3.792
23	.685	.858	1.060	1.319	1.714	2.069	2.177	2.500	2.807	3.104	3.485	3.768
24	.685	.857	1.059	1.318	1.711	2.064	2.172	2.492	2.797	3.091	3.467	3.745
25	.684	.856	1.058	1.316	1.708	2.060	2.167	2.485	2.787	3.078	3.450	3.725
26	.684	.856	1.058	1.315	1.706	2.056	2.162	2.479	2.779	3.067	3.435	3.707
27	.684	.855	1.057	1.314	1.703	2.052	2.158	2.473	2.771	3.057	3.421	3.690
28	.683	.855	1.056	1.313	1.701	2.048	2.154	2.467	2.763	3.047	3.408	3.674
29	.683	.854	1.055	1.311	1.699	2.045	2.150	2.462	2.756	3.038	3.396	3.659
30	.683	.854	1.055	1.310	1.697	2.042	2.147	2.457	2.750	3.030	3.385	3.646
40	.681	.851	1.050	1.303	1.684	2.021	2.123	2.423	2.704	2.971	3.307	3.551
50	.679	.849	1.047	1.299	1.676	2.009	2.109	2.403	2.678	2.937	3.261	3.496
60	.679	.848	1.045	1.296	1.671	2.000	2.099	2.390	2.660	2.915	3.232	3.460
80	.678	.846	1.043	1.292	1.664	1.990	2.088	2.374	2.639	2.887	3.195	3.416
100	.677	.845	1.042	1.290	1.660	1.984	2.081	2.364	2.626	2.871	3.174	3.390
1000	.675	.842	1.037	1.282	1.646	1.962	2.056	2.330	2.581	2.813	3.098	3.300
∞	.674	.841	1.036	1.282	1.645	1.960	2.054	2.326	2.576	2.807	3.091	3.291

TABLE 4

TABLE 4
χ^2 Critical Values

(Table entry is the point χ^2 with given probability p lying above it.)

Probability

$(\chi^2)^*$

df	.25	.20	.15	.10	.05	.025	.02	.01	.005	.0025	.001	.0005
							p					
1	1.32	1.64	2.07	2.71	3.84	5.02	5.41	6.63	7.88	9.14	10.83	12.12
2	2.77	3.22	3.79	4.61	5.99	7.38	7.82	9.21	10.60	11.98	13.82	15.20
3	4.11	4.64	5.32	6.25	7.81	9.35	9.84	11.34	12.84	14.32	16.27	17.73
4	5.39	5.99	6.74	7.78	9.49	11.14	11.67	13.28	14.86	16.42	18.47	20.00
5	6.63	7.29	8.12	9.24	11.07	12.83	13.39	15.09	16.75	18.39	20.51	22.11
6	7.84	8.56	9.45	10.64	12.59	14.45	15.03	16.81	18.55	20.25	22.46	24.10
7	9.04	9.80	10.75	12.02	14.07	16.01	16.62	18.48	20.28	22.04	24.32	26.02
8	10.22	11.03	12.03	13.36	15.51	17.53	18.17	20.09	21.95	23.77	26.12	27.87
9	11.39	12.24	13.29	14.68	16.92	19.02	19.68	21.67	23.59	25.46	27.88	29.67
10	12.55	13.44	14.53	15.99	18.31	20.48	21.16	23.21	25.19	27.11	29.59	31.42
11	13.70	14.63	15.77	17.28	19.68	21.92	22.62	24.72	26.76	28.73	31.26	33.14
12	14.85	15.81	16.99	18.55	21.03	23.34	24.05	26.22	28.30	30.32	32.91	34.82
13	15.98	16.98	18.20	19.81	22.36	24.74	25.47	27.69	29.82	31.88	34.53	36.48
14	17.12	18.15	19.41	21.06	23.68	26.12	26.87	29.14	31.32	33.43	36.12	38.11
15	18.25	19.31	20.60	22.31	25.00	27.49	28.26	30.58	32.80	34.95	37.70	39.72

TABLE 4

16	19.37	20.47	21.79	23.54	26.30	28.85	29.63	32.00	34.27	36.46	39.25	41.31
17	20.49	21.61	22.98	24.77	27.59	30.19	31.00	33.41	35.72	37.95	40.79	42.88
18	21.60	22.76	24.16	25.99	28.87	31.53	32.35	34.81	37.16	39.42	42.31	44.43
19	22.72	23.90	25.33	27.20	30.14	32.85	33.69	36.19	38.58	40.88	43.82	45.97
20	23.83	25.04	26.50	28.41	31.41	34.17	35.02	37.57	40.00	42.34	45.31	47.50
21	24.93	26.17	27.66	29.62	32.67	35.48	36.34	38.93	41.40	43.78	46.80	49.01
22	26.04	27.30	28.82	30.81	33.92	36.78	37.66	40.29	42.80	45.20	48.27	50.51
23	27.14	28.43	29.98	32.01	35.17	38.08	38.97	41.64	44.18	46.62	49.73	52.00
24	28.24	29.55	31.13	33.20	36.42	39.36	40.27	42.98	45.56	48.03	51.18	53.48
25	29.34	30.68	32.28	34.38	37.65	40.65	41.57	44.31	46.93	49.44	52.62	54.95
26	30.43	31.79	33.43	35.56	38.89	41.92	42.86	45.64	48.29	50.83	54.05	56.41
27	31.53	32.91	34.57	36.74	40.11	43.19	44.14	46.96	49.64	52.22	55.48	57.86
28	32.62	34.03	35.71	37.92	41.34	44.46	45.42	48.28	50.99	53.59	56.89	59.30
29	33.71	35.14	36.85	39.09	42.56	45.72	46.69	49.59	52.34	54.97	58.30	60.73
30	34.80	36.25	37.99	40.26	43.77	46.98	47.96	50.89	53.67	56.33	59.70	62.16
40	45.62	47.27	49.24	51.81	55.76	59.34	60.44	63.69	66.77	69.70	73.40	76.09
50	56.33	58.16	60.35	63.17	67.50	71.42	72.61	76.15	79.49	82.66	86.66	89.56
60	66.98	68.97	71.34	74.40	79.08	83.30	84.58	88.38	91.95	95.34	99.61	102.7
80	88.13	90.41	93.11	96.58	101.9	106.6	108.1	112.3	116.3	120.1	124.8	128.3
100	109.1	111.7	114.7	118.5	124.3	129.6	131.1	135.8	140.2	144.3	149.4	153.2

Alternative hypothesis a research hypothesis

Bar chart a graphic that displays how data fall into different categories or groups

Bell-shaped curve symmetrical, single-peaked frequency distribution

Bias the consistent underestimation or overestimation of a true value

Bimodal curve with two equal scores of highest frequency

Binomial event with only two possible outcomes

Bivariate involving two variables

Box plot (box-and-whiskers) a graphic display of data indicating symmetry and central tendency

Central Limit Theorem a rule that states that the sampling distribution of means from any population will be normal for large sample n

Chi-square a probability distribution used to test the independence of two nominal variables

Class frequency the number of observations that fall into each class interval

Class intervals categories or groups contained in frequency graphics

Coefficient of determination a measure of the proportion of each other's variability that two variables share

Confidence interval the range of values that a population parameter could take at a given level of significance

Confidence level the probability of obtaining a given result by chance

Continuous variable a variable that can be measured with whole numbers and fractional (or decimal) parts thereof

Correlation coefficient a measure of the degree to which two variables are linearly related

Critical value the value of a computed statistic used as a threshold to decide whether the null hypothesis will be rejected

Data Numerical information about variables

Degrees of freedom a parameter used to help select the critical value in some probability distributions

Dependent events events such that the outcome of one has an effect on the probability of the outcome of the other

Dependent variable a variable that is caused or influenced by another

Descriptive statistics numerical data that describe phenomena

Directional test a test of the prediction that one value is higher than another

Discrete variable a variable that can be measured only by means of whole numbers

Disjoint occurrence both outcomes unable to happen at the same time

Distribution a collection of measurements; how scores tend to be dispersed about a measurement scale

Dot plot a graphic that displays the variability in a small set of measures

Frequency histogram a graphic that displays how many measures fall into different classes

Frequency polygon a graphic presentation of frequency of a phenomenon that typically uses straight lines and points

Grouped measures a set of values that belong to the same class

Histogram a graphic presentation of frequency of a phenomenon

Independent events events such that the outcome of one has no effect on the probability of the outcome of the other

Independent variable a variable that causes, or influences, another variable

Inference conclusion about a population parameter based upon analysis of a sample statistic

Intercept the value of y at which a line crosses the vertical axis

Interquartile range set of measures lying between the lower quartile (25th percentile) and the upper quartile (75th percentile), inclusive

Interval a scale using numbers to rank order; its intervals are equal but with an arbitrary 0 point

Joint occurrence both outcomes happening simultaneously; *P(AB)*

Least squares any line- or curve-fitting model that minimizes the squared distance of data points to the line

Lower quartile (Q_1), the 25th percentile of a set of measures

Mean the sum of the measures in a distribution divided by the number of measures

Measures of central tendency descriptive measures that indicate the center of a set of values

Measures of variation descriptive measures that indicate the dispersion of a set of values

Median the middle measure in an ordered distribution

Middle quartile (Q_2), the 50th percentile of a set of measures; the median

Mode most frequent measure in a distribution

Mound-shaped curve see *bell-shaped curve*

Mutually exclusive events such that the occurrence of one precludes the occurrence of the other

Negative relationship a relationship between two variables such that when one increases, the other decreases

Nominal a scale using numbers, symbols, or names to designate different subclasses

Non-directional test a test of the prediction that two values are equal or a test that they are not equal

Non-parametric test statistical test used when assumptions about normal distribution in the population cannot be met, or when the level of measurement is ordinal or less

Normal distribution smooth bell-shaped curve symmetrical about the mean such that its shape and area obey the Empirical Rule

Null hypothesis the reverse of the research hypothesis

Ogive a graphic that displays a running total

One-tailed test a test of the prediction that one value is higher than another

Ordinal a scale using numbers or symbols to rank order; its intervals are unspecified

Outlier a data point that falls far from most other points; a score extremely divergent from the other measures of a set

Parameter a characteristic of a population

Percentile the value in an ordered set of measurements such that $P\%$ of the measures lie below that value

Pie chart a graphic that displays parts of the whole

Point estimate a number computed from a sample to represent a population parameter

Population a group of phenomena that have something in common

Positive relationship a relationship between two variables such that when one increases, the other increases, or when one decreases, the other decreases

Power the probability that a test will reject the null hypothesis when it is, in fact, false

Probability a quantitative measure of the chances for a particular outcome or outcomes

Probability distribution a smooth curve indicating the frequency distribution for a continuous random variable

Qualitative variable phenomenon measured in kind, that is, non-numerical units

Quantitative variable phenomenon measured in amounts, that is, numerical units

Random error error that occurs as a result of sampling variability

Range difference between the largest and smallest measures of a set

Ratio a scale using numbers to rank order; its intervals are equal, and the scale has an absolute 0 point

Region of acceptance the area of a probability curve in which a computed test statistic will lead to acceptance of the null hypothesis

Region of rejection the area of a probability curve in which a computed test statistic will lead to rejection of the null hypothesis

Regression a statistical procedure used to estimate the linear dependence of one or more independent variables on a dependent variable

Relative frequency the ratio of class frequency to total number of measures

Research hypothesis a prediction or expectation to be tested

Residual the vertical distance between a predicted value y and its actual value

Sample a group of members of a population selected to represent that population

Sampling distribution the distribution obtained by computing a statistic for a large number of samples drawn from the same population

Sampling variability the tendency of the same statistic computed from a number of random samples drawn from the same population to differ

Scatterplot a graphic display used to illustrate degree of correlation between two variables

Skewed a distribution displaced at one end of the scale and a tail strung out at the other end

Slope a measure of a line's slant

Standard deviation a measure of data variation; the square root of the variance

Standard error a measure of the random variability of a statistic

Standardize to convert to a z-score

Statistic a characteristic of a sample

Statistical significance the probability of obtaining a given result by chance

Statistics a branch of mathematics that describes and reasons from numerical observations; or descriptive measures of a sample

Stem-and-leaf graphic display that shows actual scores as well as distribution of classes

Symmetry a shape such that one side is the exact mirror image of the other

Systematic error the consistent underestimation or overestimation of a true value

t-distribution a probability distribution often used when the population standard deviation is not known or when the sample size is small

Tabled value the value of a computed statistic used as a threshold to decide whether the null hypothesis will be rejected

Test statistic a computed quantity used to decide hypothesis tests

Two-tailed test a test of the prediction that two values are equal, or a test that they are not equal

Type I error rejecting a null hypothesis that is, in fact, true

Type II error failing to reject a null hypothesis that is, in fact, false

Upper quartile (Q_3), the 75th percentile of a set of measures

Value a measurement or classification of a variable

Variable an observable characteristic of a phenomenon that can be measured or classified

Variance a measure of data variation; the mean of the squared deviation scores about the means of a distribution

z-score a unit of measurement obtained by subtracting the mean and dividing by the standard deviation